The Necessity of Friction

Sponsored by the Institute for Futures Studies

The Necessity of Friction

EDITED BY

Nordal Åkerman

WestviewPress

A Division of HarperCollins*Publishers*

Copyright © 1998 by Westview Press, A Division of HarperCollins Publishers, Inc.

Published in 1998 in the United States of America by Westview Press, 5500 Central Avenue, Boulder, Colorado 80301-2877, and in the United Kingdom by Westview Press, 12 Hid's Copse Road, Cumnor Hill, Oxford OX2 9JJ

A CIP catalog record for this book is available from the Library of Congress.
ISBN 0-8133-3434-9

The paper used in this publication meets the requirements of the American National Standard for Permanence of Paper for Printed Library Materials Z39.48-1984.

10   9   8   7   6   5   4   3   2   1

Contents

Foreword

Friction connotes the most basic of human conditions. The environment surrounding us presents a variety of obstacles, some that we cannot do without and others that need to be reduced or cleared away altogether to make progress possible. Virtually every form of activity, mental or physical, will be met with opposition and difficulties. In fact, friction is so fundamental and present a feature of the human predicament, and thus so taken for granted, that it is almost invisible. For a long time it has been the hidden subject of many discussions concerning development models as well as societal and technical change. It is *the* issue within a series of issues.

In history, studies of friction have been connected with names like Aristotle, Leonardo da Vinci, and Nikolaus von Cusa, among many others. Over the past two centuries, such studies have included abundant research on sliding, static, and rolling friction. Yet, tribology became a field of its own only in the 1960s, when hitherto separate efforts in physics and mechanical engineering (regarding friction), metallurgy (wear), and chemistry (lubrication) were brought together. The new interest in conservation of energy gave this work a stronger position among scholars and in politics.

Although there is an interdependence between the direct and transferred senses of the concept of friction, it is the latter, the transferred or metaphorical sense, that gets more and more important as it helps to define the possibilities of life in modern society. As friction becomes reduced in crucial ways, in one area after another, reevaluating the goals and means of the rationalist project becomes even more urgent. This is the final test of whether or not we are in control of processes we started ourselves.

Use of the concept in the transferred sense is, of course, not novel: Metaphorical references to friction can be found in various disciplines, albeit mostly under other labels (e.g., "imperfection" and "stickiness" in economics and "resistance" in psychology). But rarely does one find in any field more than a short, preliminary treatment of the concept, nor has it been employed as an instrument for analyzing the development of society. Charting a new area makes aspirations towards completeness unwise, and this work (originally published in a less edited form in Germany 1993) is rather to be seen as a first attempt at engaging the subject as a whole and as a vital force in everyone's life. I think the chapters herein will show that friction is a key that, more effectively than others, unlocks some important conditions and processes.

For me, *The Necessity of Friction* is the culmination of a twenty-year interest. Taking the idea to the Swedish Institute for Futures Studies I found a strong support and a generous grant.

Nordal Åkerman

PART ONE

Points of Departure

1

A Free-Falling Society?
Six Introductory Notes

Nordal Åkerman

Without Contraries is no Progression

—William Blake

I

Consider war.

In the face of political and nationalistic passions, safety catches give way easily enough. Suddenly, the cool planning is tested against a rapidly changing reality, making only one thing certain: Nothing will proceed precisely as decided beforehand. Life itself, in all its complexity, ambiguity, and lack of overview, will rule the battlefield.

Soldiers are told to move from A to B and in the process knock out a small enemy entrenchment at C. Should not be too difficult; but the adversary has a considerably greater force at this point than reported, and it has come out fighting. One of the transport vehicles has broken down, and in the rain and dark the maps of this unknown terrain seem slightly at odds with the world. The attackers, having become the attacked, can no longer be counted on to reinforce the troops at B. Already the somewhat bigger plan of which they were a part is flawed.

Or as Clausewitz put it: "Everything is very simple in War, but the simplest thing is difficult. These difficulties accumulate and produce a friction which no man can imagine exactly who has not seen War" (1968, p. 164).

Friction: Drunken soldiers, rain to make the path up the hill so slippery that cannons cannot be placed there, malfunctioning material, officers who

pull in different directions and discreetly sabotage the details of
the confirmed strategy, general staffs who will go to any lengths to curb
inspired mobility by local commanders, misunderstandings. ... Yet
Clausewitz teaches us that campaigns are decided not by a series of little
events but by singular bigger events, although little ones can build up
importantly. Indeed, the multitude of factors in all their combinations will
ensure that one cannot rely on set rules. Thus, as Clausewitz maintains,
the theory of war has to mean the study of war. Genius in this context is
the capacity to adapt to the unforeseeable and to exert a strong will, these
being the only means available to counter the effects of mishaps and
mistakes.

Of all the insufficiently studied factors that cause stress, locally as well
as centrally, none is more important than the speed with which events
unfold. But the quality of information is also crucial. In Clausewitz's day
the problem was one of accuracy and lack of knowledge. Currently,
however, we are faced with rather too rich a flow, necessitating computer
processing, an overstimulation that may lead to its opposite by numbing
the receiver, causing him to vacillate between preconceived guidelines and
sensory impressions.

Experience will tell commanders and soldiers alike that every effort will
be dogged by inertia, resistance, delays. What is worse, most adverse
situations will come as surprises. As Clausewitz pointed out, most will
also be related to individuals, with all their obstacles and difficulties: "This
enormous friction, which is not concentrated as in mechanics at a few points,
is therefore everywhere brought into contact with chance, and these
incidents take place upon which it was impossible to calculate, their chief
origin being chance" (1968, p. 165). Beware, lest we drown in relativities or
end up at the same station as Zeno with his "proofs" of the impossibility of
movement. Through "the haze of war" progress is still discernible, the more
so when the combatants are not of equal strength. But war is always an
untidy business. Even in so one-sided a fight as the Gulf War of 1991, among
the American soldiers lost, one in four were killed by fire from their own
side, while more died in accidents than in actual combat. Sheer will power
and meticulously planned operations cannot fully offset friction, only lessen
it, though now and then decisively.

One option, of course, is to intervene in such a way that the least possible
contact is made with the enemy. This is one of the basic tenets of American
isolationism with its notion in military strategy of the "surgical strike,"
using primarily one's air force capability to hit the opponent's strongest
bases and thereby force him into withdrawal or surrender. For the advocates
of this view, a land battle on foreign continents is to be avoided by every

means. Both Korea and Vietnam bore out such advocates regarding the risks of getting bogged down in wars that by their very duration and resulting increase in number of atrocities and casualties appeared ethically more and more dubious. What has not been borne out is the thought that air strikes alone would accomplish anything remarkable. The Gulf War gave witness to the suspicion that even "smart bombs" miss and that, however astounding the precision of many missions, the association to clean-cut surgery is more natural to the attacker in the air than to the attacked on the ground.

Each era presents its own set of obstacles to the military campaign. Historically well-known difficulties have been alcoholism, drug abuse, and ethnic strife. In our time these latter phenomena have reappeared (in connection with the former Warsaw Pact), along with such problems as unclear chains of command and parallel weapon systems (NATO). The gap between official strategy and actual planning has increased considerably, a situation that could make war truly unmanageable. For decades the West was very taken with the concept of "flexible response," which was touted as the reigning strategy, while in reality all decisionmakers assumed the immediate use of nuclear weapons should Western Europe come under attack. The option of limited war between the two super powers provided some hope of containing an armed confrontation, but this semi-official posture masked the lack of consensus regarding such crucial factors as escalation and geographical limits. Throughout *all* of history, in fact, self-deception has been a powerful instrument of the military sphere to create its own impediments. In this respect, of course, it resembles civilian life.

II

Friction is what keeps you from realizing your goals. It is what compromises all your plans, sometimes making them unrecognizable. It constitutes the divide between dream and reality. Like a translucent veil it permits you to view what you are after but often stops you from getting at it, in a tantalizingly gluey sort of way. Now and then slits in the veil will open, but not always at those points where your major effort has been directed.

Friction is that which resists, that which is inert and recalcitrant and makes almost every endeavor much harder than anticipated. It offers a series of surprising obstacles, although uncertainty is only part of the setup:

Many properties of friction are all too evident and must be expected to play their roles. Obviously, the present is what matters here: Looking ahead (utopia) demands that friction be planned out of the picture, looking backward requires that you automatically think it away, as if claiming to know exactly how things should have been.

Friction never tires. You do. Rarely is the job fully done. More often than not, the struggle starts all over again. At best, routine gives you an advantage and things will go a little easier; at worst, your Sisyphus-like task wears you down bit by bit, a slippery slope. Fighting friction, you soon realize, tends to make it worse, not unlike entropy.

Friction and inertia both act as keyholes through which you must go in order to achieve what you have set out to do. For every task, every problem, there usually is more than one keyhole, often several—though, pressed by time and opposition, you may see only one. Friction puts a premium on one-sidedness as this circumstance strengthens your will. Yet someone has to deal with the waste.

But friction is also what stops you from careening past the tangent points, ultimately ending up in outer space: It attaches you to the ground and, mysteriously, "sorts things out," albeit on a lower level than hoped for. If it stops schemes from being completely fulfilled, it also stops them from going totally awry.

As it offers the same kind of resistance to all people, friction makes us think and act in similar ways, regardless of geography and, to some extent, time. Ordinary stumbling blocks and everyday chores become, through friction and inertia, the great levelers between different cultures.

Yet without friction, there's no movement whatsoever. Nothing can get going if it cannot push off something else. A grain of sand, a difficulty, is nearly always a necessary incentive to achieve progress (the freedom gained in poetry by the regular meter). Later, en route, friction provides a perpetual contact with the world. Through it, your direction can be plotted.

It is friction that, together with the energy of motion, keeps you warm, defending life. As increasing speed brings about decreasing variety, friction comes to represent a richer world. A key concept that helps define the parameters of existence, it can thus be studied in the context of literally every human activity (and in that of some nonhuman activities as well).

All-permeating and seemingly unavoidable, friction can also be interpreted as encompassing neighboring metaphors, such as inertia. In physics, inertia connotes the properties of a single body, while friction applies to interactions between two bodies; but the scope of these concepts becomes much more ambiguous in the transferred sense, sometimes making them synonymous, sometimes subjugating the one to the other. In terms of creativity, their relationship is a question of degrees and balances; but it must be said that, where inertia rules, no invigorating friction can be expected. Generally, though, friction is the wider, more complex, and more intriguing subject, and it will be put in focus here.

III

To the proponents of the modern rationality project, every human and societal problem is solvable, every obstacle removable. Having lived all their lives, sheltered, within the long tradition of Enlightenment, they also believe in perfection. This attitude is part of the background to the tremendous advances made within science, commerce, building the welfare state in all its varieties. It also helps to explain the element of force and intolerance in the rationality project, its dependence on a continuous momentum, and its wish to throw its mesh over virtually everything, sliding into what Karl Popper calls pseudo-rationality. For the past two hundred years it has dominated the scene completely, effectively barring any alternative and giving rise to a number of violent reactions to itself. Only by remedying this project can we render the future less destructive than it appears today. And this cuts to the core of our subject matter.

When we look back through history, two particular features attract our attention. First, the aggressiveness and expansionism of most people, the will to conquer, the wild impatience with what is. The lust for power that seems stronger even than the sexual impulse. Soon enough the feeling that might is right. This has not differed over the millennia; the masks may have changed, but not the basic drive.

Second, the urge to do away with friction, that is with everything that stands in the way of progress, which is a statement bordering on the truistic and needs a qualification: Literally at any cost, with the same disregard for both man and materia. The prime vehicles in this effort have been inventions—new tools and machines that have accelerated development. In retrospect we can see that the quest has not followed a straight line. Already in the early archeological material, such as from the Sumerian civilization, one finds traces of transports by wheel and sled, and various lubricating measures—but also such friction-enhancing tools as drills, potter's wheels and ploughs adorned with stones to make them more efficient. Furthermore, throughout history many inventions have been blocked—because the social structure to make use of them was not there, because the "wrong" people promoted them, because the advances they offered did not outweigh the costs of change. Each failure has its own story and particulars, though; no overall pattern is apparent. Moreover, some inventions were lost along the way and had to be reinvented, as with several cases in ancient China (e.g. the wheel) and Egypt.

Other factors, too, cleared the way for advancement. Organization and structuring of the day and work were as important. In the thirteenth century the clock gave society an artificial rhythm distinct from the one characterized by bodily functions and the change of night and day. Indeed, mechanical time is the basis of science, said Lewis Mumford. As he also noted:

> The gain in mechanical efficiency through coordination and through the closer articulation of the day's events cannot be overestimated: while this increase cannot be measured in mere horsepower, one has only to imagine its absence today to foresee the speedy disruption and eventual collapse of our entire society. The modern industrial regime could do without coal and iron and steam easier than it could without the clock. (1963, p. 17)

Setting the scene and doing away with all the vestiges of complacency encompassed a mental adjustment on a grander scale as well. During the next few centuries, through discoveries made possible by cartography and instruments of sea navigation, the world became something other than a threatening void. Nature, if not controllable, became at least observable. Abstractions like time, money, and trade erased the limits of action.

Even more important, man became an abstraction, subject to a reductionism that discarded everything that was not quantifiable and geared toward expansion and development. This had really started in Plato's cave, where distinctions were made between the idea and its shadow, between mental entities and their representations in the material world. Later, it was immensely strengthened by Galileo and Newton. According to Galileo the world could be expressed by a mathematical formula in which all sensual properties had been removed as nonquantifiable and therefore irrelevant to rational analysis. The dichotomy between mind and body would get an even starker and more mechanistic expression in the writings of Descartes (and, farcically, in those of La Mettrie).

The gap has widened ever since. By placing the aimed-for object outside man, the frictions adhering to the individual that Clausewitz talked about could be avoided to a certain extent. Science was liberated to pursue every path of investigation that promised to enhance knowledge. Observation was divested of emotions, personal bias (one would like to believe), and the human scale. The compelling logic of development seemed to ensure advancements in one and the same direction, unhampered by any system of beliefs or values. Progress, very early on, achieved a standing of its own, not necessarily connected to man.

But a price had to be paid, and in manifold ways. As science picked up speed, reductionism increased rather than decreased, making use of ever more sophisticated knowledge to fragment that which had constituted a whole. The worrying richness of the human mind was largely pacified: The very curiosity that had fired scientists seemed bound to give results that increasingly reduced the element of surprise in human behavior. Whatever the motives of individual scientists, research came to resemble colonization, exploiting the newfound insights to tame first nature and then man. Making use of *all* knowledge was a foregone conclusion even in the absence of commercial pressure; in our time the so-called devil's doctrine means that anything that can be done *has* to be done. It is extremely hard to think of any example of a technological or scientific breakthrough that "the world" has rejected because its long-term effects would be detrimental to society.

The reason is that rationalism easily becomes less than rational. If by rational we mean an attitude that critically examines the pros and cons of arguments, abhors given "truths," and accepts that perfection can never be attained and should not be sought, then we must conclude that something, early on, went wrong in the industrialized countries (and, through their dominance, in most other countries as well). Rationalism came to stand for explicit contempt for insights and knowledge based on anything other than logic, for division of labor, and for immediate profitability and externalization of costs—all factors in the crusade against friction. One instrument has been the abridged calculation whereby only those facts that support the adopted policy will be observed. The issue of nuclear power is an example of how a belief in progress (based on emotions rather than logic) is converted into a belief in every kind of technique while the most blatant facts arguing against particular techniques are actively suppressed. Another example is, of course, the field of environment, in which costs often are hidden or denied even today. When some people propose that rationality must be retained and pseudo-rationality with its pretentions of omnipotence must be discarded, they are in effect aiming at making the calculation complete. Only in this way can alternative policies be judged fairly and a rational choice be made.

Why is it that most classic utopias, when read today, give such an artificial impression and seem so oppressive? The reason probably has to do with their all-encompassing character. No room is left for the ambiguous, the contradictory, the latent. Everything is explained, divided into its smallest parts, and put together again. All obstacles, all friction and inertia, have been brushed aside. The good will is so irrefutably good that it demands to reach every nook and cranny. The image conveyed is one of shivering coldness and spiritual poverty. We are reminded of that "iron cage of rationality" that Weber saw in the enchanted garden.

The character of the classic utopias would be a small matter, were it not for the fact that visions both reflect and guide our aspirations. Many utopias have been acted out in real life, the largest one being the planned economy. The results of this experiment show, above all, what happens when the responsibility for efficiency and care is vested with the impersonal collective instead of with the individual. The very autonomy with which leaders of planned economies operated in their attempt to surmount friction ensured that friction would rise uncontrollably at lower levels. Again, the example of the environment demonstrates the effects of clearing away checks and balances. However, when the planned economies in Eastern Europe and the Soviet Union fell it was less the counterpart of the market economy that was exposed than an important sector of the rationality project of which the market economy is another. One of the many features the two sectors have in common is that they act as if there were always some other agent to deal with those side effects that never really make it into the "rational" calculation.

The utopian impulse can also be studied at other points in history. For example, a cult of the machine appeared in the decade before the Russian Revolution. It was closely connected with a belief in progress in general and thus came to play a role in movements as diverse as Italian futurism/ fascism and Scandinavian social democracy. Emphasis was on detaching the individual from his immediate environment, turning him into a cool observer with few and crude emotions who had forsaken petit-bourgeois niceties in favor of the "modern" life. "Modern" seemed to stand for always being somewhere else with no long-term involvement, sentimental or otherwise, with any one place, a life among generalities in an advertisement. All dynamism emerged from machines—not from humans, who were expected to serve slave-like functions. And the machines themselves turned personal, whereas men and women became anonymous, interchangeable. On the horizon loomed the automated world with robots in important slots. The clearly infantile character of these dreams, so close to adventure and science-fiction books, has persisted, lately forming the emotional foundation of the search for artificial intelligence (AI) and artificial life.

Emerging somewhat later, in the second half of the 1920s, was the international style in architecture. Expounding on "less is more," it attacked the human instinct to adorn, to articulate. Now the ideals were the straight angle and the plain facade, oversized boxes of little or no variation that led to boring predictability. A hatred of the street, with all its complexities and unforeseen encounters, a rejection of its very planlessness, appeared as the motivating force. The international style parallelled the trend toward isolation in the rationality project by cutting off as many ties to the historic past, expressed in buildings and city plans, as possible. From the very start the imperialistic wish to eradicate all that had come before was there in programs and designs. Functionalism (as the international style also is called) separated working, living, and shopping, resulting in cities characterized by seemingly low-degree resistance, but only seemingly because friction rose accordingly in the form of vandalism, alcoholism, drug abuse, political apathy, many other ills. Rationalism probably never had a more revealing representation than the screaming silence of the city of functionalism. Its houses symbolize the political system it helped to shape. From the inside looking out: a clear view of a geometrical society. From the outside trying to look in: an opaque mountain of tinted glass.

But not just the masses feel shut out. Increasingly the elites discover that the disjointing mechanism triggered by systematic functionalism in all walks of life has made society impenetrable. Faced with the self-movement of technology and married to a model of development that has acquired a life of its own, they have lost all illusions of being able to control society any more, or even to understand it. This situation did not come about by chance. Man was made free from material constraints through the social engineering implemented in the modern welfare state, but the process was not matched spiritually, and perhaps could not be. For coupled with the cold expediency of rationalism, social engineering became more and more paternalistic, making the individual lonely rather than free. Denying many of his properties, treating him as an object, almost as a machine, yet asking of him to behave like a human being and a rational one at that, the elites have impoverished him and thereby themselves. This self-destructive bent in rationalism can also be studied in the reactions it causes today. Like fascism before it, fundamentalism in all major religions is a brute protest against spiritual need in a sea of abundance. It tries to reerect the inner boundaries of man by further diminishing the human potential, thus only supporting a multipolar, hedonistic world that it uses and despises at the same time. Consistent rationalism that leads to violent irrationalism, a flabby exploitative pluralism pressing the individual to fanaticism—this is the reality we have to deal with on the road from Galileo and Descartes.

When some critics of the rationality project argue that the answer to the continued abuse of inner and outer limits of the individual lies in expanding the domain of rationalism (what has been described here as making the calculation complete) while others sanction the acceptance of an irrational sphere, they miss the heart of the matter. Neither faction deals with the innate drive of rationalism to divide, to disconnect, to propel society along a one-sided logic. Neither will be able to stop its thrust into new fields, as when hitherto silent knowledge (like a butcher's skill in cutting up) is formalized into the program of a computer-guided machine (that will do the job more quickly and profitably), or natural sanctuaries are shrunk because of new gadgets (the mountain bike, the water moped), or a doctor's sound judgment and good heart (in dealing with euthanasia) are replaced with strict rules (which in this context threaten to turn the samaritan into a murderer). But even if it is doubtful indeed whether rationalism's thrust can be reversed or corrected, all means to that effect should be tried.

As noted earlier, the ambition to seriously reduce friction can be equated with progress and the quest for modernism. The prime instruments serving this ambition have been new theories and inventions resulting in historical leaps. As no invention even in the recent past has been able to offset more than a limited part of the spectrum of friction, we have not yet experienced a sea change. Although people in the rich world have seen a staggering difference in this century regarding the sheer physical effort it takes to get about, start something new etc., there is still a connection with old times. Especially so as many basic efforts have to be repeated from time to time; witness the fact that many countries now have to invest great amounts of money to renovate their infrastructures (sewer systems, bridges, water mains, etc.)—a measure all the more necessary given what seems to be a clear correlation between such investments and productivity.

But change is speeding up and with the advent of the computer, especially, and that of other barrier-breaking inventions such as hybrid DNA, we may, at least in certain sectors, envisage what has been called the frictionless slope. Actually, this vista is not so theoretical any more as it is, for example, easy to observe what the lack of friction has come to mean in the money markets, where everyday currencies are bought and sold at an average value of 1,5 trillion dollars, ten times the amount an ordinary day in 1985, and where one of the few possible remedies, the Tobin tax (on transactions), seems more or less forgotten. More and more, people realize a feeling of bewilderment as processes seem to be spinning out of control. Can anything be done about it? Can we throw spanners in the wheels of progress? Do we want to? Very little, if anything, in the history of mankind suggests that we know how to live with a reasonable amount of friction. As individuals we get to know the price of things that come too easily, but collectively we tend to shrug off any restraint. Speaking of DNA, this attitude may be part of our genetic code.

IV

The human life may seem a strange construction when viewed in terms of the individual's possibility of achieving his ends. Having seemingly overcome the first experience of friction caused by the disparity between the little child's acute, sophisticated impressions and its primitive and only slow-evolving capacity for expression, the individual has to accept that so much time is spent on preparations, such as maturing and schooling, and so relatively little in actually doing what one is aiming for. Out of an ordinary life span of seventy-five years, some thirty or forty years at most are what can be spent on any serious business with all the faculties at one's disposal. Is this nature's way of limiting the change than any one person can wreak on the world?

To one of the protagonists of Vladimir Dudintsev's *The White Clothes*, it looks that way:

> Nature pulls us away at precisely the right moment, he said after a long pause. All classics, authors and composers, all Venices and Norwegian fjords suffice for a life of average length. They sing Figaro and you sit there bored, but for those who hear it for the first time, they are delighted. If one studies this, Fedja, one will notice that everything is calculated for such a span. In that time a man can acquire experience. And when he sees that he must act and, primarily, when he sees how to act—then it's time to die! And if he manages to say to a young person: "That thing is not tasty, you don't have to try it" the young person will say: "Shut up, old man!" He wants to try it out for himself. That is, he wants to do a job already done, waste his time where it has already been wasted by someone else. It's not mere chance that nature has given us this average time-span so that we will get nowhere. It does not approve of speed. But some it orders to make leaps... (1986)

Yes, a very tangible feature of friction in the progress of history is the fact that human beings are loath to learn from each other. Every generation desperately wants to commit its own blunders. But not in all respects. No doubt we are willing to learn and copy technical achievements. Regarding consciousness and ethics, however, each one of us seems destined to start with stone-age moral concepts and work slowly upward. Most get stuck under way. The balance between technical prowess and the moral awareness instrumental in dealing with material advancements gets more and more skewed as we develop along two tracks at different speeds.

Mostly we attack phenomena in our professional world that tend to slow us down and lock us in drudgery and sameness. But the campaigns of the "little" life are no less hampered by obstacles and inertia than those of politics and large projects; in fact, much more. Where the professional life offers a certain momentum, because of the interaction of many agents and the increasing values at stake, the ordinary, everyday existence at home concerns few people with limited aspirations. Here the psychological side of friction has a much stronger sway. Never will the infuriating aspects of opposition and inertia be more keenly felt. Self-doubt and feelings of futility will not be surpressed, as in larger circumstances, but will be free to exert their full influence. A weakness of will is more easily succumbed to.

What works in the other direction is, of course, the much smaller size of the goals that makes them more attainable and thus less prone to incite frustration and the will to fail. Yet, their very smallness may obstruct the sight of greater perspectives and prevent effort. As Vincent van Gogh wrote to his brother Theo: "Don't let's forget that the little emotions are the great captains of our lives, and that we obey them without knowing it." These little emotions sometimes form a totality, a web that will not break even in the face of determined efforts — especially when reinforced by the inner enemy, the counter-will that aims at stymying our conscious wishes. For creative people, such as artists, it is a matter of routine to find ways, often strictly regular ways, around these blocks. To others the key may be operative illusions; for example, they may present to themselves a highly inflated picture of their abilities, all the while knowing that it is far from the truth.

It is considerably harder in private life than in public to put your finger on what exactly is keeping you from attaining your goals. Things seem to go to pieces all by themselves, plans to disintegrate, without any one factor sticking out. This the little girl in Gregory Bateson's "metalogue" has caught onto:

Daughter: Daddy, why do things get in a muddle?
Father: What do you mean? Things? Muddle?
Daughter: Well, people spend a lot of time tidying things, but they never seem to spend time muddling them. Things just seem to get in a muddle by themselves. And then people have to tidy them up again.
Father: But do your things get in a muddle if you don't touch them?
Daughter: No—not if nobody touches them. But if you touch them—or if anybody touches them—they get in a muddle and it's a worse muddle if it isn't me. (1972)

If little problems present themselves in never-ending succession, the opposite process also obtains. Things have a way of sorting themselves out, as in a new marriage where edges are rounded out by time. The same can be observed in new housing projects in which a great deal of initial social turbulence is followed in a few years by a functioning mode. One might think the cause to be better services, but even in projects where most amenities are present from the beginning, considerable time is required before things fall into place; yet eventually they do, in most cases. Some would say the tenants are resigning, giving up on the high profile, adjusting. Others would say they are maturely coping with the world.

Adjusting, getting along, has always been seen both as a means to achieve some comfort and as a way to increase one's efficiency. Written rules are rarely any problem for the normal individual. Unwritten ones that abound and change at an increasing pace are harder to handle, a fact that is never so clear as when immigrants try the first rungs of the ladder of social mobility. Yet the demand to melt in, unavoidable as it is, signifies the human condition more than any given time or society. Charles Affron provides this explanation:

> The failure of the Balzacian character depends upon an imperfect interpretation of the forces that control his life, on an unwillingness to make the compromises necessary for conforming to them, or on an inability to fashion his life according to them. In a sense, each novel is the account of this relationship between the human element, governed by intelligence, and the unyielding matter of reality, constituted by the forces that lie beyond man's control, the constants that must be reckoned with rather than defied or ignored. (1966, p. 5)

Whatever the circumstances in Balzac's day, in contemporary society the constants are not constant for very long. The circulation time of norms and rules, of theories guiding education and training, of ethics in relation to society and workplace, of institutions and of gadgets, has decreased drastically, causing stress and alienation for those not vital enough to cope. Postmodernism, with its ad hoc character of dissolved connections, puts a premium on alertness—not to say smartness and lack of principles, an opportunistic readiness to conform to ever-changing patterns rather than to fight for an integrated set of beliefs. This demand for constant vitality will effectively separate those who accept and thrive in a culture pandering to the superficial from those who cannot muster the strength constantly, or at all, and who also feel repelled by the lack of sincerity. Being too successful in jumping or circumventing the hurdles of this world is not without its risks, though, a theme dealt with symbolically in Goethe's *Faust* : Communicating vessels and the coupling of pressure/counterpressure are metaphors that come to mind as we observe mounting psychological problems when obstacles fall away effortlessly.

The rapid turnover in machines and things surrounding man also leads to attitudes that imply a lack of knowledge about how they work and can be fixed. Very often it is not possible to mend a broken gadget as the tools are not available or the cost of doing so is higher than the price of a new piece. For those who think that the hand has something to tell the brain and that practical know-how is essential for a well-rounded intellectual outlook, this is a sorry state of affairs. As more and more machines are fabricated by ready-made units put together in an automated fashion, the situation is hardly likely to improve.

A protest against this form of Taylorism applied to the commodity sector was heard in the 1970s when Robert Pirsig's *Zen and the Art of Motorcycle Maintenance* was published. It fast became a cult-book and helped bring about the New Age movement. Seemingly it is a novel about how to accept a reasonable amount of friction. The way to do that, according to the text, is to understand your environment, making sure that you are in control rather than being controlled. The story tells of four people on a motorcycle hike through the western part of the United States. The narrator, "I," and his friend John are juxtaposed to represent two attitudes: "I" is all for fixing things whereas John is the very opposite, a guy who uses technology without understanding it or even wanting to. John is consistently portrayed as a rather pitiful ignoramus, forever dependent on "I" to help out.

The irony of the book—and one that many readers seem to have missed—is that as "I" is at all times fixing his bike, tightening and oiling the chain, changing parts, and so on, he is also the one with constant problems. While John, doing nothing, at most encounters loose-fitting handle bars, "I" faces endless troubles with tires wearing out, funny noises from the motor, a slackening of the chain. At least subconsciously the reader must draw the conclusion that ignorance about gadgets is the wiser attitude.

Pirsig obviously has a strong case as long as he advocates *some* knowledge of technology. With marked pleasure the sermonizing "I" recounts how he has put in their place people who do not understand the simple things around them. But the impression one gets from this philosophical tract, thinly disguised as a poetical novel, is the notion that one has to know it all to be in charge of one's life. This unfortunately cannot be a realistic program for our times; the best we can opt for is to prevent a total cop-out, especially among young people, which would mean treating the whole of modern existence as a kind of machine, not to be figured out but expected to produce happiness.

The authors of cult-books are often mental bullies who have found a few notable truths and thereafter ask to guide your life. Pirsig is no exception. But it should be noted that his overall answer leads in a direction exactly opposite that recommended by the New Age movement. To Pirsig the solution to the malaise of our times is not occultism, right-brain exercises, and the like, but an expansion of the field commanded by rationality. Though the overtones of the narrative clearly could lead to other conclusions, the long biking trek ends up as a middle-of-the-road, welfare-oriented policy of common sense that normally embraces a totally different, quite devout view of technology. Pirsig's dilemma seems even more pungent today, more than two decades after the publication of his tract.

V

A skeptical attitude toward the development brought about by one-sided rationalism need not refer to any Luddite sentiments. Neither is the crucial point the number of machines or their size (although numbers and size play a role in alienation). Even the monster Caterpillar tractors used for strip mining can be seen as extensions of the human arm. As Mumford described the situation:

> Machines have developed out of a complex of non-organic agents for converting energy, for performing work, for enlarging the mechanical or sensory capacities of the human body, or for reducing to a measurable order and regularity the processes of life. The automaton is the last step in a process that began with the use of one part or another of the human body as a tool.(1963, p. 9–10)

But when the robot is given a wider capacity and is used in more and more areas, at some point the border at which quantity is transformed into quality will be crossed. Quite a few people probably believe that this border will never be crossed, as it hasn't been already. We are still awaiting the much-heralded invasion by robots—in factories (which they have only partly penetrated), in homes, in communications. Obviously some techniques are discounted decades too soon (recall, for example, the proposed neutron bomb that caused a lively debate in the early 1960s, twenty-five years before its existence), though eventually they do appear. Take a look at your living room; it probably has not changed much in the last forty years, if at all. The TV set has acquired colors, but that's about it. In the rest of your house you will find perhaps a more sophisticated telephone, a microwave oven, a few remarkable children's toys—and a PC. This clearly will change very rapidly now. Soon almost everything in your home will be affected by computerization and miniaturization. From the moment you open the front door by voice activator or retina scanner you will handle tools and machines strikingly different from those you find in today's world. Very many of them will have a small built-in computer that, in a limited sense, will "think" for you and imperceptibly decide what *can* be done, what *should* be done, and thus what you *want* to do. On this massive scale such technology will dominate the environment. Only the naive can believe that it will not change man's view of himself and force him to adapt.

Another aspect of the present development is polarization. The basic idea of the computer age is reductionism/compression. Why, then, is the IT society so full of nonsense and never-ending chitchat (as on the Internet)? The reason is that the new technology clearly acts as a bifurcation, sending the elites off to their secluded and intense activities while the vast majority of people are pushed into an area where entertainment reigns in various forms. On all levels the instrument used promises that one will always come to the right place, get the right answer, elegantly avoiding all traps and potholes. But as all knowledge seekers (e.g., those searching for books in a database) will attest, getting the wrong answer can sometimes lead to encounters and insights of great value. What are the hidden costs incurred by a technology that in every instance takes us to the goal we believe we are looking for? Will not the intoxicating speed with which we can get anything we want soon make us impatient with those things and people who cannot keep up?

Yet, what's really crucial will be observed as the computer gains importance in creative work and as research on artificial intelligence (AI) opens up new vistas. Previously, intellectual capacity always had its final limits in the constraints of the body. Knowledge has been won through and with the body, in spite of the many gigantic projects that have exceeded the human scale altogether. If we look closely enough, we will still find a connecting line, a continuation of the primitive building that muscles achieved.

But with the help of the computer the brain is now lifted out of its bodily restrictions, the tactile base of its knowledge fades away, and soon we are sliding down the frictionless slope that so far has been only a daydream. The computer already enables us to simulate a Vermeer painting, to make up photos that even experts find it hard to expose as artificial, to perform at present hundreds of millions computations in a second and soon billions, in minute detail reproduce on the screen a human cell. Personal, computer-edited newspapers will ensure that we are never again bothered with views or items in which we have shown little or no interest in the past. With the computer glove and voice-command computer we will even be rid of that remaining obstacle—having to use a symbolic language. In cyberspace/virtual reality we will move *into* the computer to become part of a world that we ourselves manipulate totally, where we will meet and interact with people from the other side of the planet as if we were in the same room. As reality blurs and blends with fantasy, ethical constraints fall by the roadside. Like the anti-hero of Albert Camus' *The Stranger*, we will increasingly experience the outside world as a mass of disconnected entities, shimmering in the haze. It seems that the tradition of Plato, Galileo and Descartes has finally been redeemed.

To the critics of AI research as it is practiced today, these fears are entirely unfounded. Justifiably, they feel that the pretentions of this sector in science are inversely proportional to the achievements made so far. Most of these critics believe that the computer is a smart machine, more and more fast and sophisticated but nevertheless a tool under human control that society needs in order to develop. Although they acknowledge and accept the enormous power of the computer, they not only mock some of the more pathetic experiments of AI research but maintain, on theoretical grounds, that such research *cannot* succeed in its attempts to create really intelligent, man-like machines.

What then is the computer in relation to us? In the words of Sherry Turkle one can talk about its between and betwixt character: "As a mind that is not quite a mind it provokes questions about mind itself." In spite of what many say, it is not neutral. Its importance does not stem from the context into which it is placed; rather, it forces this context to change or it creates its own context. It presupposes that the user will adapt to a rationalistic, nonassociative way of thinking. And, contrary to what is constantly proclaimed, it has further strengthened centralist tendencies in society, thereby increasing the vulnerability of modern life.

As a couple of enthusiasts have pointed out, the strengths and weaknesses of the computer have been the very opposite of what was expected at the start:

> AI has presented us with a paradox: all the elegantly structured symbolic artefacts that we think make us most human, such as mathematics and logic and the ability to splice genes or infer underground geological facts, are what computers handle best, because the more highly structured knowledge is, the easier it is for us to codify it for computer use. On the other hand, getting around in the real world is not a highly structured task—the average house pet can manage that easily, but machines cannot. This is not to say that they won't ever be able to: it's a statement about affairs at the moment. (Feigenbaum and McCorduck, 1983, p. 57)

There is probably no one in the world who can beat the fastest computer in a game of chess. When ruling world chess champion Garry Kasparov met Deep Blue, a computer with the capacity of 200 million computations a second, he managed with some difficulty to get a win. Pitted against Deep Blue II in May 1997, he lost and was so confounded by its new properties that he publicly voiced suspicions of foul play (implying that the computer at crucial draws was supported by another chess master). The moment was of great symbolic significance: For the first time on this level a human being gave vent to his frustration over the discomforting feeling of having been surpassed by a machine. Yet it has been an unsurmountable problem to program a computer to act adequately playing with a child or visiting a restaurant. Many encouraging examples can be cited of the body's tremendous abilities, some of which we never give a second thought or even know about—such as the eye's capacity for computing edges, a process that engages 100,000 cells in the cortex, of which each is capable of as many operations per second as the fastest computers today.

Still, the advantage of humans seems only relative. New leaps of intensive development will give the computer a raw power that may equate with certain bodily functions. Through parallel processing and neural networks—perhaps in the future coupled whith laser beams for storing and retrieving data on the molecular level—complexity is advancing dramatically. As if incredible speed were not enough, AI research now appears to have found a more fruitful avenue of progress (perhaps indicated by the much wider acceptance of expert systems in their later generations), concentrating on the things that computers do best. Probably many delays have been caused by efforts to imitate all of man. In fact, from the perspective of machine logic, innumerable human attributes are irrelevant and indeed disturbing. Our need for walks in the woods is not shared by the computer. Perhaps it will suffice to imitate a limited amount of mind structures and properties of man, yet to create a culture in which he becomes a prisoner. We have very much eased such a development through putting a premium on the quantifiable at the expense of the intuitive and through our consistent reductionist effort, which in science is a tool but in politics and culture causes a process of impoverishment. Hence the cross-bar that AI research has to leap may be resting at a considerably lower notch than generally imagined.

Is this a reductionist argument, given the apparent presupposition that man can be replaced by something as simple as lightning-fast computations? Only seemingly. We are building a computer's world that is shaped to suit machines and machine logic. It is built by so many and therefore invisible hands that it appears irresistible. Should this technique meet with success, it will be because a number of key structures in human thought and behavior have been identified and copied. It is then the technique that is reductionist, not the critique against it. When "the new mystics" (adherents of a school of thought within AI research) maintain that man is something more than the sum of the knowledge and abilities with which the other schools are trying to supply the computer, they are certainly right. But what does that really mean? Being "more than the sum" is hardly a guarantee that man will stay on top forever.

The kind of machine culture discussed here is not built along one track only. It is the combined effect of many layers of development trends that have to be calculated if one is to get a clear picture of what the future of computers might mean and what possible social harm AI research might entail even if it does not make it over the cross-bar. Among these trends, some of the most important are

- the efforts to give mind a geography, whereby different categories of thought can be localized to different parts of the brain;
- the bionic man; more and more body parts can be replaced artificially which opens up the possibility of
- the cyborg (cybernetical organism), the growing together of man and machine with guidance via miniaturized computers;
- computers that are "humanized" through a sort of associative thinking with elements of randomized idiosyncrasies (while man is expected to conform in just the opposite direction);
- recombinant DNA-techniques, with their limitless perspectives of leaps in the development of species and the creation of quite new species without niches in the ecology established over time;
- increased control of the individual as the dark side of laudable projects (like a bio-sensor to be worn by the professional driver—only by him?— to measure his possible fatigue and level of intoxication);
- a society adapted to the demands of technology without debate and without real decisions.

To each one of these points some will say that it is not a bad thing in itself or that it does not matter too much. Who can argue against saving lives? But what we have to ask is whether we are rapidly setting the conditions for a mechanization of the soul and an absurd separation of emotions and intellect. Will the total effect of these and other trends bring about such a radically new role for man that in fifty or a hundred years we will have become what *today* we would not accept as human beings? The counterargument, that those who embarked on the first train were expected by some people to die from the experience, is of no use. What concerns us here is not only a change in the perception of reality but a change that is objectively verifiable, fundamental, and complex.

The computer has the overwhelming force of everyday existence. It can store an almost limitless amount of facts, which it can retrieve in no time at all (as humans cannot do). It suffers no inhibitions, no embarrassment, no *idées fixes*, almost no friction. With its self-correcting software it travels in its own space. In a short time it has colonized the world (or we have allowed it to do so) in a way that is not really reversible. Smart young people are fond of saying that you can always pull the plug, but this is not true: As long as our paychecks depend on the plug staying in, staying in it will. Regarding our awareness of the great and growing vulnerability in society caused by the structure that computers foster it has to be said that the only really invulnerable actor will be the computer itself, whose well-being we give a higher priority than anything else.

The relationship between man and computer cannot help but change when the latter has been given properties that will eventually make it far more creative than it is today (and then is invited to help develop - past itself—new generations). In all probability, we will partly withdraw; sometimes be ashamed of our lack of competitiveness and then experience feelings of obsolescence. Yet as long as we can identify a role for ourselves, we will cooperate. By the same token we can also refute the most forceful argument belittling the adjustments we will see—namely, the argument that, as computers never can have intentionality, we will always be in command. For we will be the ones who supply the intentionality for as long as it is needed—that is, until the program of the computer is so rich and varied that a "will" arises. That will is still with the operators. But the symbiosis is steadily changing as computers gain in strength. In one area after another, we truly cannot ride counter to the computer's assessment because the input of information is so torrential that only machines can handle it. Thus, the issue is not whether the computer can be given some kind of intentionality but, rather, whether we are building systems that make us abdicate our own intentionality or make it the tool of the machines.

This is not to accept the claims that when a cell is visually reproduced on a computer screen "artificial life" has been born or that the entire population of computers in the world is the equivalent of the brain with all its synapses. Analogies like these do not tell us how to get from the screen to the street (if, indeed, that is the goal). Nor should we be overly impressed by nonprogrammed anomalous surprises that occur as two computers "talk" to each other. But the muddled thinking behind these claims helps to blur the contours of reality, increasing the general feeling of uncertainty that eats away at the core of accountable, participatory democracy. Again, even when they fail, attempts at doing away with friction can have wide social repercussions.

Let us face the development squarely. The final aim of the efforts to create artificial intelligence is to make man superfluous. The scientists in the field do not hide this ambition. Although their papers often make an astonishing impression—starting with a few pages on the difficulties involved in making a robot walk across the room followed by many pages containing speculations in a purely science fiction spirit—they are clear about what they are up to. As expected, they find few moral problems in their work; one of them has remarked that they see no difference in principle between a brain based on carbon atoms and one based on silicon atoms. Only too aware of the imperfections of humanity they stand ready to help something "better" into this world.

And how about people in this apocalyptic perspective? Perhaps a twelve-year-old in Turkle's child populations has the future in his hand:

> When there are computers that are just as smart as people, the computers will do a lot of the jobs, but there will still be things for the people to do. They will run the restaurants, taste the food, and they will be the ones who will love each other, have families and love each other. I guess they'll still be the only ones who go to church. (1984, p. 62)

A reservation, this time for all humans. But ten, twelve or fifteen billion people use a lot of energy and produce as much garbage and problems. Soon enough the very concept of man will appear irrational in that rational world he himself has made up. And that will be the end of the story. Something is of course bound to remain, for a while, in the form of a machine instruction and a program. But this something cannot be anything else than what is captured in digital-1 and digital-0.

VI

As described by the Marquis de Condorcet, the Enlightenment philosopher, human improvement is a relentless struggle to balance the power of nature. And because man is weak and subjected to her laws, he can modify them only through the combined and persistent efforts of many generations. Today we can observe that radical changes are brought about by much lesser work. Now we ruefully contemplate the effects of battles won only too well.

Nevertheless, de Condorcet had a point, in that friction always seems to gain the final edge in the life of the individual. Through vitality and ideas she can keep ahead for a while but, in the long run, resistance will present her with an eternal opposing force. Not so for the society. There, new people constantly offer to lend their weight, and the impressive series of victories has at least the appearance of being definitive. There, the line of change is always stronger than the line of conservation. Projects attain a momentum that precludes much deliberation and consultation. The pull is cumulative and accelerating. Suddenly the shackles fall off on several fronts at the same time.

To the individual the leaps are of some concern, but what dominates his consciousness is rather the incremental character of development, which also gives it its element of coercion. Every attempt at seriously raising the possibility of charting another course will seem, on both the individual and political levels, unrealistic if not nonsensical. On both levels an economic squeeze and various forms of private and public contracts will ensure that no one rocks the boat. Modest alternatives will be scoffed at as uninteresting, radical ones viewed as steps into the dark. It follows that any analysis that delineates the consequences of present trends is less than welcome. If reality is only bearable as it is, then its codification will seem unwarranted and painful, and will elicit aggressive denials.

So we adapt, almost frantically, to the demands put to us by a society systematically developed as if it were meant eventually to be taken over by thinking machines. Indeed, a machine-like behavior is increasingly approved of: Feelings of shame and guilt are actively surpressed as outmoded; concern for others is ridiculed in everyday business, and appreciated only when associated with cloying sentimentality as in films; success is now a value in itself, regardless of the reason and effort, or lack thereof, behind it; even the manic repetition is considered normal and indeed interesting. We read about painters who do not paint, sculptors who do not sculpt, but let others do the job. Gesture is everything. Intellectuals declare themselves to be working for peace, yet do not hesitate to write lyrically about plays and films that idolize murderers and rapists. In a process that started for real in the 1960s, men and women feel "liberated" as they manage to overcome inhibitions, the primary mark of civilization. Parents refuse to act the role of parents, refuse to offer the opposition every young person needs to tackle and conquer in order to grow up. A new type of personality is coming to the fore, one in which surface and core coincide. It could also be described as a Peer Gynt type, excellent at getting around problems by seductive talk, proficient at keeping up with the powers that be, but shallow and utterly unreliable.

This heavy shift of emphasis in the culture of society toward the surface of things is not matched by a similar process in the sciences. On the contrary, unlike the humanities, where tendencies like these are manifested now and then, the sciences are exhibiting increasing specialization and more limited contacts with neighboring disciplines. What the elite levels have in common, though, is a set of conditions in which friction is of less and less importance. At the other end of the social spectrum the situation is quite different. The desertification of the poorer sectors of the cityscape, the neo-Dickensian quality of life for many in the rich countries, caused by conscious political decisions and a change in technology that means a huge net loss of jobs, is but one example of violently increasing friction. The shiny, glossy, sleek facades of the postmodern society are not for the underprivileged, who, by way of contrast, feel more lost and without hope than their counterparts before the building of the welfare state. Enhancing this trend, multinational companies simply refuse to pay taxes in their countries of origin, establishing themselves in taxfree havens around the world, thereby hollowing out the state on which they nevertheless make excessive demands for new pieces of infrastructure and on which they rely completely in times of adversity. As for the Third World, this division between rich and poor is many, many times starker, reflecting a development model that was thrust upon it and which is maintaining the local elites on islands of resistance-free living in a sea of poverty and inertia.

Thus society is in a state of disequilibrium, with different sectors facing different futures. Integration of countries and markets will probably reinforce the picture. The elements working toward a more mature handling of the difficulties and obstacles of this existence, an acceptance of friction and a readiness to live through it and with it, are few and not very strong. Within religions, especially the Christian ones, the emphasis is on dealing with problems rather than avoiding them. Unfortunately, the dominant force in the churches now is a fundamentalism that no doubt accepts friction but also sharpens it and ties it to aggressive proselytizing. Its reaction to pluralism is confrontation. The seasoned, literate religious force that effectively could balance the powers of modernism is not in the ascendant.

Women, compared to men, have always had quite a different attitude vis-à-vis friction. Traditional tasks like childbearing and homemaking defy the impulse to speed up. Less inclined toward the utopian, the messianic, women make use of what there is, in an ad hoc way, so to speak, and they often recognize the value added to life by friction. As they now enter and share spheres where men have been predominant, female priorities may come to humanize the world. The pessimist would argue that the reason women accepted friction is that they had to, whereas now they will surely adopt the male system of values, including the lust for power. Maybe their role as childbearers, potential or actual, will insulate against that risk.

The key word in our discussion of the struggle to reduce friction is *vulnerability*. Attempts to jump-start modernity—that is, to circumvent necessary phases of history—often backfire, leading to the very opposite of what was intended (as the Shah of Iran learned to his, and our, cost). Once inside the new life, we eliminate as many as possible of those obstacles that gave us the opportunity to pause, reconsider, and opt for another route. The formula "just in time" ensures that no human consideration is allowed to stand in the way of "progress". We accept slums but no slack. Cutting the marginals, establishing new links, reducing time waste—all this in reality brings society to circumstances in which everything has to function without the slightest hitch if catastrophes are not to follow. As on the expressway, where a small mistake by a single driver can cause an accident involving more than a hundred cars, the kind of modernity we have chosen asks each one of us, as members of society, to accept a risk calculation that is absurd and is getting progressively worse. The wish to create a perfect society peopled by all but perfect humans is not a rational one.

An arrow shot from its string is hard to stop. On the road from Plato and Galileo we have reached a point where the frictionless slope is no longer a figment of imagination but really, if dimly, in sight. *All at once* is now a possibility from tailor-made cars to tailor-made humans. By scraping the intermediate stages, the inertia-riddled transportation stretches, we leap from friction into fiction. As "the spirit of time" makes us accept the nonlinear, nonhierarchical perspective, subject to ambiguity and uncertainty, on the brink of chaos, new theories within the natural sciences neatly follow suit, in that order, dreamlike, as if the world rests in "as if." The speed is dizzying, and society—humankind—is heading for a free fall. It is all according to our wish.

References

Affron, Charles. *Patterns of Failure in* La Comedie Humaine. London, 1966.
Bateson, Gregory. *Steps to An Ecology of Mind.* New York, 1972.
Clausewitz, Carl von. *On War.* London: Penguin,1968.
Dowson, Duncan. *History of Tribology.* London, 1979.
Dudintsev, Vladimir. *Belye odezjdy (The White Clothes).* Leningrad, 1986.
Feigenbaum, Edward, and McCorduck, Pamela. *The Fifth Generation.* Reading, Mass., 1983.
Forschheimer, Claire, Ed. *Knowledge and Communication in the Computer Age.* Proceedings from a symposium at the University of Linköping/Tema T. Sweden, 1988.

Manuel, Frank E. *The Prophets of Paris*. New York, 1965.
Minsky, Marvin. *Robotics*. New York, 1985.
Mumford, Lewis. *Technics and Civilization*. New York, 1963.
Pirsig, Robert. *Zen and the Art of Motorcycle Maintenance*. London, 1974.
Turkle, Sherry. *The Second Self*. New York, 1984.

PART TWO

On the Battlefield

2

Friction and Warfare

Chris Donnelly

Everything in war is very simple, but the simplest thing is very difficult. These difficulties accumulate and produce a friction which no man can imagine exactly who has not seen war. ... Countless minor incidents—the kind one can never really foresee—combine to lower the general level of performance. So that one always falls short of the intended goal. ... The military machine ... is basically very simple and easy to manage. But we should bear in mind that none of its components is of one piece: each part is composed of individuals, every one of whom retains his potential of friction. ... A battalion is made up of individuals, the least important of whom may chance to delay things and sometimes make them go wrong. ...

Fog can prevent the enemy from being seen in time, a gun from firing when it should, a report from reaching the commanding officer. Rain can prevent a battalion from arriving, make another late by keeping it not three but eight hours on the march, ruin a cavalry charge by bogging the horses down in mud, etc.

Action in war is like moving in a resistant element. Just as the simplest and most natural movement, walking, cannot easily be performed in water, so in war it is difficult for normal efforts to achieve even moderate results.

Thus wrote Carl von Clausewitz in his work *On War*, the most famous and influential study of war ever written. It is so precisely because the author did not choose, as do most writers on this subject, to address simply strategy, tactics, weaponry, and the like. Rather, drawing on a strong Prussian military-intellectual background and his experiences in action with the Prussian and Russian armies during the Napoleonic wars, he addressed the essence of war as a social phenomenon, and the essence of why, as such, it differed from all other types of social phenomena. It is this feature that makes Clausewitz's work not just of historical interest but useful today in that it helps us understand war itself—both as a tool of policy (i.e., the continuation of a national policy by means of violence) and as warfare, or which the most characteristic process is *friction*[1].

Clausewitz maintained that no nation's strategy could be effective if it did not take account of the nature of war and did not see clearly the capabilities and limitations of this political tool. Nor could any military theory or commander's plan be of value if it did not take into account the "variable factors" of warfare—that is, the uncertainty of all information, the importance of morale, and the unpredictability of the enemy. War is dangerous; it is a realm of suffering, confusion, exhaustion, and fear. These are the factors that create what Clausewitz termed *friction*, thus introducing the word into our modern usage as a permanent element of the social machine.

Clausewitz's genius did not lie in his invention of new ways of waging war. Nor was he even the first to identify the principles on which his thesis on war was built. Rather, what set him apart from other strategists was the sum of his analysis of war as a social phenomenon, and his analysis of the nature of battles and of the armies that fight them. By subjecting war to a rigorous academic study, he was able to establish principles in a logical and rational sequence that anyone with experience of war would instantly recognize as valid, and that could be taught rather than just learned through painful experience.

His identification of friction as the element that most distinguished war from other social activities was based on the premise that the principle element of warfare is not weaponry but man. It was this perception that led him to bring to our attention the vital aspect of the concept of *morale*, the first conceptual pillar on which his thesis is based. In the final analysis, it is the destruction of moral rather than physical strength at which war is directed. "War," he wrote, "is a trial of moral and physical forces by means of the latter. ... All war assumes human weakness and seeks to exploit it." Moral factors he considered the ultimate determinants in war, but how to evaluate such an intangible factor and its interaction with physical factors? It is in his study of this issue that von Clausewitz is still relevant today. As he himself pointed out, the issue is not unique to the military but common to many activities. In the years since he wrote his famous work, the development of technology and of industrial economies have become factors to add to the equation, but neither negate the value of his analysis. "It is *friction*," wrote Clausewitz, "that distinguishes real war from war on paper." And (technology and economies being roughly equitable) it is the morale, of the army and of the national population, that will determine the outcome of wars.

The second concept on which Clausewitz builds his thesis is a recognition that war is a tool of policy—war is to be used as an objective implement of policy when it suits a given nation's needs. Only this approach will enable the national leadership to plan effectively for the development of armed forces and a "military doctrine" (a concept for raising, maintaining, and using a military machine) based on the existing social and economic factors of the nation. To the Anglo-Saxons liberals of the nineteenth and early twentieth centuries, this observation was evidence of militaristic cynicism. They viewed war as an undesirable interruption of peaceful progress—something regrettably necessary at times, but to be dismissed from the attention once completed—rather than as an aberration. The study of war was not really accepted as a respectable academic discipline. They believed that armed forces raised for war should be dispersed in peacetime, and that war was evidence of the result of the *failure* of the political system, not a permanent and integral element of it. Because the Anglo-Saxon nations were victorious in the major wars of the twentieth century, this thinking has patterned the entire military philosophy of the West and dictated the shape of most of its armed forces. However, the Clausewitzean concept was adopted enthusiastically by Marx and Lenin, and it is that fact, and not liberal democratic principles, which shaped the thinking of the military theorists. On this basis they built and used the Red Army, developing a military system that later came to shape the military systems of all the Warsaw Pact countries. These military systems and concepts are significantly different from those in the West, and that difference, at the level of military organization and training, is quite evident both in the attitude toward friction and the methods used to overcome its effect, as well as in the attitude toward war as a tool of policy.

This is not to say that the impact of friction in war is unappreciated in European armies, or that, today, Clausewitz's genius is unrecognized. Far from it. However, in national attitudes toward and preparation for war, the differences between Western and Eastern European nations are quite startling, as are the philosophies on which their armies are organized. All armies, if they are to be successful, must evolve ways to cope with friction; it is the differences in their means of doing so that are of particular interest.

Friction in Battle

The preparation for and reaction to friction as a phenomenon differ in various military systems. Its basic causes are physical, psychological, and emotional factors that affect the individual and the group in different ways. First and foremost, consideration of the impact of friction forcibly and constantly reminds us that the primary weapon of war is man, so it is the impact of "stress factors" creating friction on the man in battle that must be our first topic of study. It is not possible to list these primary causes of friction on the soldier—the "stress factors"—in any order of priority, because their importance will naturally differ depending upon individuals and circumstances. However, the body of available Western literature on the subject, the recollections of veterans, and this author's own *very* limited experience of peacetime soldiering as a reservist call forth the factors in a certain sequence: fatigue, incompetence, fear, bereavement, and pain.

Let us first consider fatigue. It is a fact that most battles begin with soldiers and officers already so tired that, in peacetime, most of us would hesitate to accept a lift into town in any car they happened to be driving, so concerned would we be over their state of fatigue. During the Battle of Waterloo in 1815, many of the British regiments had marched over 80 km in two days with only two brief halts of a few hours each; when committed to the line of battle many were so exhausted that they could barely stay awake. By the time his tank battalion first engaged the attacking Syrian division on the Golan Heights in 1973, Lt. Col. Avigdor Kahalani had had no rest for thirty-six hours, having been dragged from mending the roof of his house into a mad frenzy to mobilize and move his unit up to the front.

These are not unusual instances but typical ones, and the mechanization of warfare has not diminished the two principal causes of fatigue—deprivation of sleep and extreme and prolonged physical exertion—rather it has increased them. The Roman legionnaire normally carried a pack weighing 28 kg, though it was sometimes as heavy as 45 kg. The French World War I *poilu* carried a 40-kg pack, plus his weapon. British infantry in the Falklands in 1982 marched with over 50 kg of pack, weapons, and ammunition, and often went into action carrying 35–40 kg of equipment. Today, medical studies have demonstrated the importance of daily sleep. Yet the ever-increasing speed of warfare and the spread of night-vision devices make it ever more difficult for the soldier to be sure of getting the four hours of continuous sleep that is considered essential to maintaining a minimum level of alertness over periods of more than three to four days.

If the plight of the exhausted infantry soldier in battle is bad, the impact of fatigue on the commander is far worse. On him rests the burden of thinking, planning, and reacting; on him lies the fate of his subordinates; and on him depend their lives as well as his own. The higher in rank a commander is, the more the system of command seems to conspire to load him with work. He may escape the physical labor of his soldiers, but his workload is no less arduous, because it is much more demanding in terms of alert, imaginative, or logical thought. It is hard enough to fire a rifle when tired; it is far more difficult to operate complex machinery, and even more difficult still to cope with making a plan in the face of the massive influx of information, and given the speed of movement and reaction required in a high-technology war. The march of technological progress is increasing both the causes and the impact of friction on the battlefield. The impact of any single individual's mistake can be far greater in a war today than was the case in Clausewitz's time.

In fact, the anxiety generated by responsibility is a growing cause of fatigue in modern war. The AA missile or radar operator may not face the physical strain imposed by the infantryman's pack or the increasingly heavy shells the artilleryman must load, but the strain imposed by the responsibility of his task is just as real as a source of tiredness and therefore of friction. The worry of command has always been a prime cause of stress. Officers have always been haunted by their mistakes when these have cost lives, and fear of making mistakes has paralyzed many a commander. But today the private soldier or sailor or airman will also, in many armies, carry for long periods a hideous burden of responsibility that his counterpart of an earlier age did not. This is a burden that can sap the will and even destroy the mind.

To the trials of physical exertion, lack of sleep, and anxiety caused by responsibility, the burdens of weather and climate add their load. We are usually aware of such burdens only when they come in real extremes; consider the effect of winter on Napoleon's and Hitler's invasions of Russia or the dangers of heat exhaustion in the Burmese jungles of 1943 or during Operation Desert Storm in 1991. However, soldiers have to struggle with weather in every campaign. Clausewitz was much concerned with its effects. For sailors, it is always the *main* threat; the actions of the enemy are a poor second. The unpredictability of weather is as great a cause of friction as its actual effects. Here, too, modern warfare has not improved matters. Modern equipment may be more capable of operation in bad weather conditions, but it also forces upon the soldier, sailor, or airman the necessity to operate in conditions that, in an earlier age, would have guaranteed a rest. The crew of a modern all-weather combat aircraft flying several sorties a day is not better off than the crew of a World War II Lancaster bomber tied to the airfield by a rain storm. Mud bogs down heavy armored vehicles just as effectively as it did the cavalry of Clausewitz's day, inflicting fatigue on both man and machine and frustration on the commander—just as efficient a creator of friction as it ever was. Yet another cause of fatigue, fear, and friction in battle is noise. The din of an ancient or medieval battlefield would have been limited to the shouts and screams of men and horses and the clanging of weapons as in a demonic smithy. But the advent of fire-arms and cannon raised the impact of noise to the point where it is a much more significant factor in fatigue. The din of vehicles and the ever louder crash of weaponry have increased not only the volume but the harmful quality of noise inflicted on the soldier, despite the availability of ear de-fenders. The constant noise alone of an aerial or artillery bombardment is sufficient to prevent logical thought, quite apart from the possible lethal effects of the shells themselves.

Pain and death are usually taken to be the most serious causes of friction in war, yet in fact, it is the fear of pain and the death of others which, at the basic human level, are most pernicious in wearing away the all-important morale. For most soldiers in effective and cohesive armies, the causes of fear, in order of *increasing* importance, seem to be death, disfigurement, failure, and disgrace. Soldiers are more often ebilitated by fear of disfigurement than by fear of death, and their motivation is more often provided by fear of showing cowardice or fear of letting their mates down than by fear of death or injury. Many commanders will readily risk death rather than failure and disgrace in the eyes of their peers. This contradictory effect of fear produces a corresponding friction. Fear can paralyze the soldier, sailor or airman into inaction, and it can goad the commander into rash or ill-considered action, both disastrous to the military objective. The shock effects of the horrors of battle have always been with us. But however horrific the wounds inflicted by medieval weapons were, they occured within the experience of anyone of the time—cuts, gashes, crushings, dismembered limbs. They were injuries inflicted by clearly understood causes in a restricted area. The impact of black powder weapons increased this horror, but only a little. The majority of wounds were still inflicted by simple weapons at restricted ranges. A cannon ball, to be sure, inflicted terrible wounds, but still within a certain limited and easily comprehended framework. However, in the last century, the injuries that can befall the soldier have escalated out of all proportion to the past, and the horror induced by fear of modern weapons' effects has likewise increased.

In one area, of course, the reverse is true. Hygiene and medical assistance on the modern battlefield have come to mean that disease is no longer the main killer, as was the case up to the end of the eighteenth century. In the abortive British expedition to Walcheren in 1809, 7,000 men died of disease and twice as many were invalided by the same cause, compared to the 217 killed in action—an extreme example, perhaps, but still an instructive one. Men weakened by colds, dysentery, and diarrhea or plagued by infections are more readily subject to stress factors that induce friction than are the healthy. Reliability of food supplies and provision of a nutritious diet are similarly important factors. They contribute not only to basic health and strength but also to the all-important factor of morale. Conversely, malnutrition saps energy and morale. In addition, the preparation and eating of food are very important factors in creating the social bonding between soldiers essential for team work and group cohesion. So, in this sense, the lot of today's soldier can be said to have been improved by the advance of technology.

More important than this in overcoming fear of battle itself is the knowledge that effective medical assistance will be available to the wounded. In the days before anesthesia, the soldier's suffering often began in earnest on the operating table. And in pre-antibiotic times, no matter how good the surgery, infection carryied off a large proportion of the injured whose survival would now be taken for granted. Even so, many soldiers injured in battle today can still face long waits before they reach the life-saving field hospitals (up to thirty-six hours in the case of some British troops wounded in the Falklands War of 1982), due to isolation, the flow of battle, or inclement weather that prevents evacuation. Furthermore, there is the common phenomenon of an injured soldier being attended to by numerous companions, some motivated by concern for his welfare, others by the fact that attending the injured provides a good excuse for not exposing oneself to further combat. Either way, the unit's strength is lowered, necessitating orders to leave the wounded with only the very minimum of attention until after the battle. Furthermore, the experiences of Western nations in recent years have been of limited war, when casualty care facilities more than adequatly met the challenge posed by casualties. In a really major, high-speed war, however, this would not be the case, and fear of the results of injury would once again add to the friction of war.

Fear of injury or death and the discomfort of climate and weather can easily be appreciated as causes of stress and friction, but there is one even more obvious stress factor that is in fact often overlooked—bereavement. Soldiers in war often become extremely fond of one another. Their shared experiences and suffering create a bond of affection and love that is one of the principal causes of military effectiveness in good armies and, indeed, is perhaps the main reason why so many men remember their days in the army with such affection. When the going gets really tough, one of the basic motivating factors that keeps soldiers fighting is that they fight for each other. The emotional shock that the death of a close companion can have on an individual, therefore, is enormous. The effect of bereavement in a family is well understood. The impact on the surviving partner in a marriage, for example, is itself one of the commonest causes of premature death, and at the very least emotional instability is to be expected as an inevitable outcome. The same, then, is true of soldiers bereaved of their comrades in battle. The emotional shock blurs judgment and erodes analytical capability and objectiveness. It engenders passivity and despair in some and aggressiveness in others. Even in the best of circumstances, soldiers bereaved become less predictable and usually less easy to control, a fact that greatly aggravates the friction that, in turn, so debilitates command.

Friction and Intelligence

The aforementioned causes of stress apply, at least in some measure, to the commander as well as to the serving soldier. Depending on his rank and appointment, the commander or staff officer may share the discomforts of war with his men. However, the burden of command brings with it its own particular stresses. Some unique, some akin to those of the soldier but an order of magnitude greater. All of these greatly increase the potential for friction. Moreover, by the very fact of his being in command, or in a position of staff responsibility, the commander/officer has far greater opportunity to generate friction for others than does the ordinary soldier. The greater the span of that command, the greater the capacity to generate (or reduce) the friction of war.

The first burden—fatigue—I have already referred to. It is one of the most important but also most easily overlooked factors. Likewise, bereavement—which deprives an officer of his balanced judgment, leading him, for example, to an unwise reaction for the sake of revenge—can have wider-reaching consequences than for the soldier, whose loss of judgment usually leads only to a personal or local disaster.

The commander, in particular, faces stresses that affect him as a specific function of command. The first is fear of the result of his actions, and of making a mistake. Fear of being responsible for the death of others has led many a commander into a passivity which, at the end of the day, has been far more disastrous than any activity could have been. Fear of another kind is fear of failure, and of resulting disgrace in the eyes of one's peers. Such fear is usually a spur to action rather than passivity, but it is not necessarily logical action. History abounds with examples of unnecessary losses caused by rival commanders seeking to outdo one another or seeking glory for themselves, no matter what the cost. In both cases, the emotion of fear destroys balanced judgment and adds to the accumulating friction—the greater the commander's span of command, the greater the problem.

A particular problem of command, much alluded to by Clausewitz, is the problem of information. Throughout the centuries, military commanders have complained about the inadequacy of their information—insufficient, inaccurate, unverifiable, or, worst of all, contradictory. On the basis of this information they must make plans and issue the orders on which success or failure depends. This indeed is the focal point of friction in war; it is here that all the factors mentioned earlier accumulate to form the commander's most difficult problem: the conversion of raw pieces of disconnected information in all their imperfection into a comprehensible picture of what is going on. This is the process of intelligence.

It is the task of intelligence officers to evaluate the information, corroborate reports, and provide a composite picture of the battle or war appropriate to the needs of the commander. Even in peacetime the production of intelligence has never been an easy task. In war, with the enemy doing everything possible to prevent the formation of that composite picture and, indeed, to mislead the intelligence officers by providing false information, intelligence becomes the fulcrum on which success in battle pivots. Weather, fatigue, fear, and enemy action that physically disrupts the source or passage of information all serve to reduce the quality of information reaching headquarters. Most important of all, the soldier's view of the battlefield, once battle is joined, becomes very narrow. The brain, assaulted by the shock of war—death and destruction, noise, immense physical and mental overload, excessive responsibility—responds by shutting out all unnecessary activity and restricting the span of attention to the immediately necessary. This reaction is equally applicable to all involved. The infantry soldier, whose "tunnel vision" gives him a view of the battlefield restricted to his immediate physical environment, will have no idea of the course of the battalion's battle. In turn, the battalion commander, preoccupied with his task, will have little interest in the activities of his flanking battalions except insofar as it immediately and directly affects his own battle. No matter how effective the technical links, information in battle virtually never flows "sideways"—that is, to one's neighbor. It only goes "up" to higher command levels or "down" to subordinates. Even this movement is uncertain. One would have a hard time persuading a soldier who is fighting for his very life to take any time at all to pass on to his commander a view of what is happening. Terrain, smoke, weather, fear, and physical destruction all hinder the passage of information even by those troops whose specific task is to provide it. The commander and staff, struggling with the problems of battle, plus the problems of command, find it very difficult to make time to pass back to higher headquarters (HQ) their view of what is happening even when they are succeeding. When they are failing in their task, the "difficult" often becomes "impossible." Thus the commander is left blind to events affecting the constituent elements of his command.

This problem has become steadily more complex with the change in the nature of war since Clausewitz's time. In his day the short range of weapons meant that battles were fought in relatively restricted spaces with high densities of troops. Troops, even artillery, more than a few hundred yards from the point of fighting could have no effect on the battle. With the improvement of muskets in the mid-nineteenth century, which increased their killing range from 100 to over 500 m, battles began to extend in space—compare the battles of Napoleon with those of the American Civil War. The introduction of rapid firing weapons at the turn of the century made dense deployment of troops potentially catastrophic, as many World War I battles demonstrated. Consequently, as battle extended in space, the commander's capacity to understand what was happening became less and less, and he was obliged to establish a hierarchical chain of information and command. Herein lies today's problem.

At the Battle of Waterloo, the Duke of Wellington's forces (and Napoleon's too, for that matter) were extended over a front of 3 km and to a depth of 300–700 m. It was quite possible for him to gather information and issue orders personally by riding from one end of the battlefield to another. Only the smoke obscured his view of virtually the whole field of carnage, at its greatest about 3 km by 2 km, on which some 70,000 men and a third as many horses on each side were fighting. Today, the same area would be covered by the weapons of a mechanized battalion in attack and a company battle team in defense—that is, by 400–600 men and 100–150 men plus armored vehicles and so on, respectively. Nor is today's battle likely to remain static, rather, it tends to be or become very fluid, moving at the speed of a tank or infantry carrier across the ground. Thus, not only is it becoming more and more difficult to see what is going on and to pass back the information, it is also more physically difficult for the intelligence staff and commanders to work because of the constant threat to their personal safety as well as the constant need for movement and relocation.

I stress this point especially because current portrayals of war, in films and in the news media, often give a very false impression of the reality of the situation. In fact, this is nothing new. The depiction of battle as the nineteenth century officer cadet would have seen it in the heroic paintings adorning his college or in the literature available to him would have given him no real inkling of the reality of the chaos. It is this fact that gives

Clausewitz's work such unique importance. Even so, translating his intellectual statement into personal feelings would be beyond the capacity of any inexperienced reader. The fact is that most soldiers go into battle unprepared for the reality of the event.

Today, both soldiers and civilians have a view of war shaped by Hollywood and the TV news camera. Hollywood is usually concerned with telling a story and presenting an impressive image. A film director would find it very difficult to depict the emptiness of a modern battlefield or to represent the true sense of confusion. It is not impossible to do so; it has been done, but rarely. Much more often, battle is presented as an ordered affair in wich the commander does know what is going on, can make a plan and implement it, and defeats the enemy neatly in under two hours' screentime. Moreover, even the most horrific of modern special effects cannot create the horror of warfare. In the movies, men fall with a clean hole in the head or chest, blood is spattered on a nearby wall, bodies fall neatly, quiet and still, and stay dead. The reality is very different. For every dead casualty, there are three to four men who are noisily and distressingly injured. The wounds inflicted by today's high-velocity weapons are truly ghastly. Shellfire dismembers bodies, scattering fragments far and wide. No cinema or training film, however graphic, can reproduce the stench of disemboweled or burning bodies that so assaults the senses and is itself such a cause of stress in the combatant.

The conveyance of information is similarly distorted. Great advances have been made in recent years in the technology designed to acquire and transmit information. Spy satellites, aircraft and drones, electronic eavesdropping, radar, infrared radiation, image intensification, thermal imaging, and so on—all provide a potential for more information. Of course, this has improved the commander's capacity to know what is going on; but when the enemy has more or less the same facilities, coupled with the same capacity for deception, it can be argued that the commander's position is not necessarily bettered, just altered, and made indeed more complex and difficult. The problem, of course, is an *excess* of information, which overloads the systems intended to deal with it. In fact, too much information is just as difficult for the staff to convert into *intelligence* as too little information. Intelligence, moreover, is useless if it is not delivered to those who need it in time for it to be acted upon.

If one takes as an example the Gulf War of 1991. Press reports brought to the attention of the world the latest technological marvels of information gathering, target acquisition, precision bombing, and self-guided cruise missiles. It is certainly true that the U.S. airborne information and control systems provided a hitherto unparalleled view of the battlefield, which accounted for much of the Allied success with minimum casualties. However, it is worth noting that this was a war in which the bulk of the best intelligence and high-tech military assets of the Western world, especially of the United States, were employed against a Middle Eastern state with a population and a military technology about the same as those of Belgium. Even so, there was a serious intelligence misassessment of Iraq's strength before the war, a considerable difficulty in finding and hitting the obsolete "Scud" missiles, and a very serious problem in processing and using as tactical intelligence the information gathered by the various high-tech sensors once the battle had begun. This despite the fact that the Iraqis put up no effective resistance at all on the ground and made no serious effort to target the intelligence-gathering assets or HQs themselves. Had the coalition force faced a competent and active enemy, and one equipped with a similarly effective information system, the outcome of the Gulf War might have been different. As it is, that war provides a good example of friction in battle occurring even in circumstances where the enemy does not add to the problems.

Recent years have witnessed, both in Western armies and in the public, a very strong tendency to attribute great war-winning properties almost uniquely to high technology, leading to a reliance on technological superiority to achieve victory at low cost. It cannot be denied that technology *can* be a war winner if there is a large and unbridgeable technological gap, as in the Gulf War. This, of course, is not new; recall Hillaire Belloc's pronouncement that the imperialist British need not fear the Hottentots, as "we have the Maxim gun and they have not." However, among industrialized nations, the technological gap is historically ephemeral. It takes only a very short time for nations to learn new technology and to copy it. In such instances, only *surprise use* of technology will have the desired effect, and this is remarkably difficult to achieve—specifically, because the need to keep a revolutionary new system or weapon secret precludes practicing with it. This in turn seriously hampers efforts to plan how best to employ the technology, and to train soldiers in its use. As the Gulf War showed, a lot of time and effort are required to effectively integrate new weapons and systems into the orchestra of battle. Of course, the Gulf War also demonstrated, of course, the truth of Clausewitz's dictums on the overriding importance of man and morale. All authorities are agreed that, had the Iraqis been motivated and trained to fight, even their limited technology would have been adequate to cause serious grief to the coalition.

It is possible that future intelligence systems (i.e., information gathering, processing, and dissemination) will be perfected, and that fully automated weaponry and command and control systems, by eliminating man from the operating loop, will come to remove the friction of war. However, the evidence does not point that way. If the enemy is similarly equipped, then his first target for destruction, electronic blinding or deception will be the other side's intelligence and command and control systems (C^3I in military jargon). No nation, not even the United States, has large resources of highly accurate weaponry because of its cost, and production rates are slow. Once all the cruise missiles have been used up, then any major war will entail a return to a slogging match, with friction in all its many forms reemerging— perhaps to an even greater extent than ever if the command and control systems on which the army relied have been destroyed and it has to fall back on more primitive but reliable methods. Alas, technology by itself is unlikely to solve the problem of friction, far from it. Given the increasing complexity of equipment and systems, their vulnerability to failure, and the demands made on the operator for mental and physical acuity, it is likely in the long run that technology, unless very well managed, will actually increase the friction of war and battle.

No review of the causes of friction in war would be complete without some reference to plain incompetence by commanders. Incompetence by a soldier, unless he is operating a crucial piece of high-tech equipment, is likely to be catastrophic only for himself and his immediate colleagues. However, incompetence on the part of a commander can bring catastrophe to all his subordinates and even his entire country. The annals of history are full of military disasters—not just defeats, but defeats that can be laid at the door of incompetent command. In such instances, the commander or the command system actively creates friction instead of reducing it and working through it as a good commander and staff would.

A good case can be made that the causes of command incompetence are inherent in the military system; that the very methods armies use to overcome the problems of battle, in combination with their peacetime leadership selection system for officers, almost guarantee built-in character flaws in most commanders. However unpalatable it may be to soldiers, there is at least some truth in this theory, and it can be used to explain, at least in some measure, the differing national and cultural attitudes and values that exist. For example. Britain and the United States share some general cultural values, and there is a remarkable similarity in the assessed causes of their military disasters. These I would list as follows:

1. A tendency to underestimate the enemy, and to give too little attention to obtaining adequate intelligence.
2. An inclination to reject or ignore information that is unpalatable or that clashes with preconceived ideas.

3. Indecision or avoidance of difficult decisions.
4. Obstinate persistence in sticking to a plan and ignoring evidence of its failure.
5. An inability to wage war quickly and a corresponding failure to exploit success.
6. A belief in the use of brute force, and a failure to appreciate the value of surprise and deception.
7. Adherence to outmoded traditions, an unwillingness to change to meet new challenges, and an inability to learn from mistakes.

Armies, in victory or in defeat, have a tremendous capacity to create myths about themselves and the enemy. The British and U.S. armies today frequently ascribe their World War II successes to skill and cunning, or technological advantage, and ascribe Soviet successes against the same adversary—Nazi Germany—to superior numbers alone. Yet then General Eisenhower, when asked in 1945 what he considered to be the three most important war-winning weapons, replied "The bulldozer, the Dakota, and the 10-tonne truck!" Compared to the Russian military, the Anglo-American armies made very little use of skilled maneuver, and fought what were largely battles of attrition by land and air. How many of our generals displayed the faults previously enumerated? How many of our victories were due more to success in the factory than to cleverness on the field of battle? It is the painful task of the military historian to keep us intellectually honest in our evaluation of past wars so that we can learn the real lessons. But it does seem that this intellectual honesty is constantly under threat from human nature, political expediency, and nationalist fervor. The less we insist on rigorous intellectual honesty, however painful, in the evaluation of our military actions, the more we store up future causes of friction in war, and the more we condemn ourselves to relearn old lessons.

Coping with Friction

I have outlined just some of the factors that create friction in war. It is not possible, within the scope of a short chapter, to compare in detail the various methods by which the world's armies have attempted to come to terms with this phenomenon. In any case, no one has ever been able to identify any single feature of friction, or of the means adopted to deal with it, that is unique to any nation's army. Rather, all armies *in principle* face the same problems, but to varying degrees. All armies also use largely the same methods *in principle*, but, again, to varying degrees. Of course, the degrees can vary so greatly that they result in very different answers to the complex equation constituting the military way of dealing with friction in battle.

The baseline for determining how an army copes with the problem of friction starts with the society it springs from. This society will have beliefs and values that the armed forces are charged to protect, and that in large measure shape the philosophy of the same armed forces. To this factor must be added the effects of geography, economics, and historical experience, which will determine the kind of army the nation thinks it must possess to ensure its sovereignty.

For an illustration of this point, we need look no further than NATO. Of the sixteen members of this Alliance, all but Iceland possess armed forces. All members are committed to the common defense of certain shared values of democracy and self-determination. Yet within that Alliance, each nation chooses its own form of military organization, and the military systems chosen are dramatically different. As a simple example, consider the U.K. and Norway, countries that have a higher than average level of mutual respect and friendship, and undoubtedly share many common values. British troops exercise every year in Norway, training alongside local forces and sharing a tradition of excellent cooperation and collaboration. Yet the military organization of the two countries could hardly be more different.

Geographically, Norway is a large and very extended country, though its population is small. To defend it requires large forces, but it is able to spend only a limited amount. Its history, particularly its painful experience of World War II, has convinced it of the overwhelming importance of engaging all of its people in support of a national defense policy and involving them in the armed forces. Accordingly, Norway has chosen a mass rapid mobilization system for territorial defense, a small core of regular forces that trains conscripts for mobilization into a large army in wartime. Such an army is limited in the range of its equipment, and the short training time available limits the sophistication of the tactics its troops can carry out. However, the inhospitable terrain and climate, and the local capacity to cope with that, combine with the military system to produce a very credible national defensive capacity.

The U.K. does not have Norway's recent history of invasion, sitting as it does behind a barrier of sea. Its sea power, and latterly airpower, have ensured national sovereignty. Its army has a longstanding history of colonial war and has only intermittently been engaged in large-scale mainland European conflict. The U.K. has been able to maintain, relative to its population, a small but fully professional armed force with a very limited territorial reserve capacity and no rapid mass mobilization capacity at all.

Hence British servicemen have long experience and training to fight in many parts of the world.

Both armies face friction in war, but the factors that create this friction are bound to differ in degree, as are the two armies' means of dealing with it. Training or fighting in North Norway in winter, the British troops find that the climate creates far more friction for them than for the Norwegian unit on their flank, despite the latter's shorter training time. Numerous historical examples would likewise indicate that, when both units come under intense bombardment or fire for the first time, the more experienced soldiers, and those trained longer together as a fighting team, tend to withstand the effects with less stress and a lesser "coefficient of friction"; thus they are more capable of returning fire or taking other action. Conventional wisdom would have it that the soldier fighting to defend his homeland would show more determination than the one fighting on foreign soil. However, whilst this may well be an influential factor in some cases, it is by no means demonstrable historically that this will always be the case. Examples to the contrary are legion. Thus the British and Norwegian soldier, fighting side by side in the same war, would expect to face common problems somewhat differently. Different qualities—or perhaps qualities in different measure—will make for their success or failure.

Despite the different military problems that armies face, they share certain features in common, by which they seek to mitigate the effects of friction. The most fundamental of these features are physical fitness, discipline, and drills.

1. Physical fitness is essential to cope with fatigue, that great producer of friction. This is one of the basic reasons why the bulk of armies is formed by *young* men.
2. For most people, what is demanded of them in battle is unnatural and in direct contradiction to their normal human instincts. Thus discipline is required. It demands that the individual suppress his natural inclinations and obey commands that may put him at risk for the benefit of the group.
3. The first victim of the shock or fear of combat is logical and inventive thought. Under the stress of battle and fatigue, the human capability proven to survive the longest is the capability to carry out physical or mental procedures that are so well rehearsed as to have become automatic. The better a military system is at devising the appropriate drills and inculcating these in its soldiers before combat, the more successful that army will be at reducing the friction of war at that level. This principle applies as much to the musket drill of the eighteenth

century Prussian infantryman learned in countless repetitions on the drill square, before facing the threat of death from a cavalry charge at 50 m, as to the modern intelligence officer exhausted by lack of sleep, attempting to complete his analysis under immense pressure of time, a complex military situation, and the threat of missile or gas attack on his HQ. Both benefit from the use of a well-rehearsed procedure involving algorithms and flow charts to guide their actions and give them the nature of a drill.

Team spirit goes under many names, but it is perhaps the most important feature of all, reflecting the fact that soldiers will fight for the sake of their close comrades far longer than for any nationalist or ideological cause. This behavior is due to a complex interaction of factors, including the fear of being seen to "let the side down," love of one's comrades, the fact that morale is more easily maintained in adversity when one is in a group rather than alone, the sense of self-preservation that creates a flock instinct, and the desire to shine in the eyes of those whose respect and love one values.

Of course, soldiers achieve team spirit by working together over a long period and, particularly, by sharing adversity. Team spirit is the quality most difficult to develop in short-term conscript armies during peacetime, yet also the most practical justification for brutal discipline in an army, in that the sense of adversity this discipline creates (i.e., fear of arbitrary and cruel punishment) substitutes for the adversity produced by the enemy in battle.

Friction and Command

The second level of requirements for dealing with friction in war concerns the selection and training of the commanders, and the developing of a staff system to enable them to do their work. Here, the above principles apply, just as they do to the fighting soldier, but there are additional requirements which are often difficult to reconcile, or which often bring with them undesirable side effects. These requirements are most often lumped together under the term *leadership*, itself an intangible, composed of various essential qualities. For example:

1. Courage is a desirable asset in any soldier, of course, but it is essential in a commander. This quality refers not only to the commander's bravery in battle (so as to inspire his men to follow his example) but

also to the much more difficult virtue that enables the commander under exceptional circumstances to challenge or disobey a standing instruction or the orders of a superior. This virtue contradicts the concept of discipline, and it requires a most difficult balance if it is not to be disruptive. U.K. and U.S. armies call courage "initiative." Its potential for creating friction is greater than virtually anything else.

2. Intelligence and analytical capability (in the human rather than military sense) are equally essential, increasingly so as an officer rises in rank. Yet here, too, there is a contradiction, because a young officer's experience and the discipline to which he is subjected can prevent the development of these qualities. They are prerequisites for the successful exercise of the kind of courage and initiative referred to earlier. Education, formal or otherwise, is necessary to enable them to be developed and used effectively.

3. Command authority is usually seen as the first requirement of a commander. It is certainly the most immediate. Yet if not coupled with the foregoing qualities, it will be counterproductive. In some, command authority is an innate quality; in others, it is a learned quality. It can also have undesirable side effects: Love of command for its own sake, and the adoption of a dictatorial style, can easily frustrate the development of the other essential qualities in subordinates— and in battle it is upon his subordinates' effectiveness that the commander's success depends.

Lack of such qualities is, of course, closely connected with the causes of failure referred to earlier. The challenge is to devise a military system that can balance these contradictory requirements and thereby deal with the problems of the battlefield so as to reduce debilitating and potentially disastrous friction to a minimum. It is made more difficult by the fact that the qualities required by society of a soldier in peacetime—namely, good behavior and socially acceptable manners—are not the qualities most valuable in battle, where the capacity to generate well-organized violence and destruction are all-important. The soldier and officer skilled in peacekeeping or low-intensity operations (which require great restraint in the use of violence over a long period) will not necessarily be successful in a general war. The qualities that make a man a good company or battalion commander will not necessarily make him a good general.

It is the function of the military system to create in peacetime an organization and training system that, first, trains and educates its soldiers and officers appropriately; second, provides them with the requisite equipment; third, creates a structured organization for men and equipment to operate within; and fourth, devises effective tactics, operations, and strategy based on the limitations of men and material, the requirements of the military task to be accomplished, and the nature of the enemy.

Friction, or rather the potential for it, is lurking at every point on this path, waiting to reduce the efficiency of the military machine. Overcoming this friction is the constant struggle of a military system as it strives for perfection. This is difficult enough to do even in peacetime. In war, even if under great pressure we succeed in exhausting the resources of friction, the enemy will quickly replenish these by his disruptive activity.

"No plan survives contact with the enemy," runs a well-known military proverb. Yet as soon as contact is joined with the enemy, preoccupations of the combatants and physical isolation from the commanders make it difficult to implement changes in the plan. The units are left to fight on their own. Infantry advancing on the Somme in 1916 were soon completely out of contact with their higher command and artillery, even though the objective were only a very few kilometers deep, and the advance at a walking pace at best. Today, the means of communication are far better than the field telephone of 1916, but the capabilities for enormously high rates of advance and the consequent fluidity of battle make even higher demands on the command and control systems. The modern battle group commander who fights a competent enemy is just as likely as his 1916 counterpart to find himself without meaningful instructions from his superior and without the capability to call down assistance. In other words, the fog of war will descend, and the unit will be fighting on its own. Friction will have done its work.

Ironically, we need friction—and in particular, the friction provided by an enemy in war—to perfect a military system and keep it "up to scratch" as society develops. Armies need wars just to stay effective. It is through the very act of overcoming the inevitable friction of war that armies can truly be said to have achieved improvement. Friction in this light is like the devil in *Faust*, "that power which seeks forever evil and does forever good."

Philosophizing, however, can be allowed only a very limited place in a chapter devoted to the practical subject of war and friction. Friction is an inevitable and, indeed, essential ingredient of war, and all armies strive to overcome it. It is possible to attempt this goal by purely pragmatic measures, based entirely on experience. Yet as technology and politics change the face of battle and immeasurably complicate warfare, it is difficult to envisage the successful coping with friction of an army that has not made some attempt to institutionalize the problem and apply academic analysis to attempt its solution. Such analysis has been the function of military theorists since the time of Sun Tzu and Xerxes, and it is what Clausewitz did best of all.

As I previously noted, Clausewitz's philosophy did not suit the Anglo-Saxon approach to war. In the Anglo-Saxon world, war is seen as an undesirable aberration, as evidence of diplomatic failure, as something to be accepted and coped with for sure, but ideally to be avoided at almost (but not quite) all cost. This ideal is shared by, or has since 1945 been spread to, all of the liberal democracies that we loosely refer to as "the West." But Clausewitz's ideas eventually found acceptance in his native Prussia, and then in tsarist Russia. They were also closely read by Marx and Engels, and later particularly by Lenin, and became part of the fundamentals of Marxist-Leninist ideology. As such, they were incorporated into the thinking of the Soviet state and the Soviet Armed Forces. The Soviet Armed Forces became the real heirs of Clausewitz. The Soviet military system made the most strenuous attempt to realize his teaching, both in its broader concept, i.e., the use of war as a tool of policy—and in its application to battle and armies—building a military system that attempted the systematic exploitation of what were seen as the principles of armed conflict.

The Soviet Approach to Friction

The Soviet military system did not claim perfection, but it claimed to strive for it. The Soviet Union may have disintegrated under social and economic pressures, but many aspects of its military system live on in the constituent elements of the former Soviet Armed Forces, in particular in the Russian Armed Forces which claim the bulk of the Soviet military inheritance. It is not my purpose here to analyze the reasons for the failure of the Soviet system. Undoubtedly, excessive investment in military might was partly to blame and overcoming the problems created by this excessive militarization is one of the most serious obstacles to the development of Russia and the other ex-Soviet republics as viable democracies with free-market economies. However, in our obsession with the fall of the U.S.S.R. let us not overlook those elements of the Soviet system that actually worked well, in particular of its military system. This was not a perfect system by any means; its excessive influence on the economy, its lack of adaptation to a changing society, its obsession with preparing for war with the West, and its consequent inability to adapt to war in Afghanistan all proved fatal to society. Yet the Soviet Army did an excellent job of developing the theory and practice of a military system per se, and there is a great deal we can learn from this, inasmuch as the system strove to minimize friction for itself and to maximize and exploit friction in the enemy's ranks. The Russian Army of today continues this heritage, and will undoubtedly develop it further if Russian society overcomes its current turbulent problems.

The Soviet, and today Russian, attitude toward battlefield stress has been characterized by the following qualitites:

1. widespread recognition of the causes and effects of stress on the individual and of the threat that the resultant friction poses to the viability of the unit in combat;
2. realization of the value of exploiting the effect of stress and friction on the enemy;
3. methodical study of all aspects of the military system designed to identify points at which the individual and the system are particularly vulnerable to stress and friction;
4. coordinated and sustained efforts to incorporate into training and tactics (a) such measures as would serve to decrease the soldiers' own vulnerability and (b) such measures as would increase to critical levels the stress and friction imposed on the enemy at crucial points and times in the battle.

The key to understanding the Soviet and Russian military systems is an appreciation of the function and importance of military doctrine.

There is no real equivalent in the British and U.S. armies, and indeed in most other Western armies, of such a military doctrine. Western armies generally pride themselves on inculcating initiative in their officers and soldiers, and on the high degree of independence of action allowed to commanders at all levels. They tend to assume that a military doctrine stifles initiative, destroys independence of thought (and therefore of action,) and leads to stereotype. Although even the Russians would agree that there is an element of truth in this assessment, it is far from a fair definition of their concept of a military doctrine in which they have such pride and on which they still rely so heavily. On the other hand, the Russians often deride Western "reliance on native wit" (which is how they translate the Western concept of initiative), considering it a poor substitute for a carefully worked out and detailed plan.

The Russians see their military doctrine as the accumulated wisdom and experience of generations of competent Soviet soldiers. This body of knowledge is constantly being amended by experimentation, exercises, and a meticulous academic and practical study of past wars. It is impossible not to be impressed by the volume and quality of detailed analyses of the problems of the battlefield that are carried out by Russian military scholars. These analyses and debates—which regularly appeared in the open Soviet press and, despite all that has happened since 1989, continue to appear in the Russian military press—represent a sustained effort to keep military doctrine up to date in all its strategic and tactical aspects.

The second distinguishing element of the Soviet, now Russian, military system is its General Staff, which concept sprang from the Prussian and Russian Imperial General Staff traditions and was developed in the infant Red Army by B. M. Shaposhnikov. It is a staff within a staff, an elite of specially chosen officers whose task is to mastermind military development, plan the improvement of the military system, and tend to the doctrine. The officers of the General Staff are the key workers who translate doctrine into action: organization, equipment, training, and strategy. They control military intelligence (the GRU), the educational establishments, and their work is used a basis for directing the course of the military industrial complex. They are the brain of the army and, as such, impose their will on the sinews, overcoming by their domination much of the friction generated between elements in a less dictatorial military system. For example, their power and cohesiveness as a group have for many years largely overcome the interservice rivalry that creates so much institutional friction in many Western armed forces.

The combination of a doctrinal system and a General Staff is a most powerful one. Its disadvantage is that it is so powerful that it cannot be challenged. If friction is to institutions what natural selection is to evolution in nature, then the General Staff and doctrinal system, in attempting to create a smooth running military machine as devoid of friction as possible, has removed friction from the national level of military command. This it did in the U.S.S.R., first, by establishing an identity of interest with the ruling Communist Party and the defense industrial complex and, second, by monopolizing all military authority and expertise. There was no civil control of the military in the U.S.S.R.: no friction between institutional bodies striving for power and authority as is commonly found between foreign and defense ministries in Western democracies, and no struggle for finance between defense and social spending as is inevitable in democratic market economies in peacetime. The Party was concerned only with ensuring the army's loyalty and readiness to carry out its tasks. The U.S.S.R. never developed any civilian expertise in defense and military affairs whatsoever. There were (and still largely are) no expert parliamentarians, academics, or journalists to challenge military policy decisions; no forum for public or even informed private discussion; no financial responsibility or accountability; indeed, no fiscal limits at all in the absence of a hard currency and the lack of any concept of a budget. What we have seen, in other words, is the maintenance of a wartime military system since 1945, and the results have been disastrous for the national economy. This is what the *absence* of friction can do, whereas the friction with which Western democracies live permanently is taken for granted, annoying and "wasteful" though it may at times appear to be.

Nor is the Soviet system free from criticism when it comes to purely military organizational measures. There are many in the Russian army today who blame the General Staff's preoccupation with preparing for a war against the West as self-sustaining, as part of the Party's implacable ideological and nationalist perception of a Western threat. This preoccupation has prevented the General Staff from allowing the armed forces to change in keeping with the changes in society, from learning the lessons of the war in Afghanistan, and, today, from adapting to the changes brought about by the end of the Cold War.

These criticisms are clearly justified, and there is a very important message in recent events in the former U.S.S.R.. But let us not forget that the former Soviet military system, in the form of the Russian Army and Russian General Staff, has survived the fall of the U.S.S.R.. It has also maintained its basic principles, but under a rejuvenated command team who intend to use the system to rebuild the Russian Army and reestablish Russia as a great power. As a system intended purely for generating combat power in wartime, the Soviet system was first class, and it is this status that the General Staff seeks to preserve for the Russian Army of tomorrow. The example of their military organization and philosophy will continue to be with us as long as there is a Russian General Staff and a Russian Army. It is important to point out here that, although the Soviet and Russian concept of war remains firmly wedded to the primacy of the offensive, this does not make the system aggressive, nor do I mean to imply that the Russian General Staff seeks to wage war on the West in pursuance of a militaristic aim. I most definitely do not. The system is too firmly imbued with the Clausewitzian concept of war as a tool of policy—with the teaching that war is only a last resort, to be used when one has no alternative or, if given a choice, only when success is absolutely assured.

It is also worth noting that the members of the Soviet General Staff considered their concept of a "friction-free" national doctrinal system to be the world's most effective system for generating military power, and that they extended this concept to the Warsaw Pact. In war the Pact armies were to be directed to do a specific task in a defined area as an integral part of a Soviet theater operation controlled by the Soviet General Staff. There was never an independent multinational Warsaw Pact command and control system such as exists in NATO. By molding the military systems of the Warsaw Pact member nations in the Soviet model, and by denying to these nations any truly independent military system, they considered that they had achieved a very effective alliance, one that would move to a single military command, and whose members, being organized on comparable lines, would mesh together in action with as little friction as possible.

NATO, on the other hand, was considered a most awkward kind of alliance. The Russians saw it as frittering away its economic and technological advantages because it allowed its members to develop independent military doctrines, independent national command systems, different methods of training, no common tactics and unnecessarily diverse types of equipment. Such conditions, in Russian eyes, were bound to create friction and inhibit effectiveness in war.

In any event, it is NATO that has survived and the Warsaw Pact that has failed; but the latter failed the test of peace, not of war. There are good grounds for thinking that, had it been a question of war between East and West, and not peace, the outcome might have been very different. As far as efficiency *within* a military system is concerned, then, the Soviet and Russian military system is still worth examining because of the very positive military lessons we can learn from it, especially those relevant to the subject of this chapter.

The upbringing and training of Soviet soldiers and officers, battle tactics, organization and equipment, weapons design and procurement, even relations among the army, the Communist Party, and Soviet society, all these aspects of the Soviet military system were molded by Soviet military doctrine. The result was a military system characterized by (1) an enormous degree of integration and coordination between all its constituent elements; (2) total consistency between all its elements regarding goals and objectives of the system and means of implementing plans; (3) a widespread awareness of the interdependence and interreaction of all elements of the system; and (4) a widespread conviction in the value of the system and in its ideological correctness, coupled with a pragmatic approach to the search for solutions to problems of the battlefield.

For many years, Soviet military research was not above criticism: It was severely hampered by an excess of ideological orientation and a fear of overfrank admissions of failure, and it was concentrated almost exclusively on the Soviet experience, which is of warfare of a particular scale and style. However, within these limitations, the Russians studied and analyzed battle from all perspectives, and their conclusions concerning the effect of stress on soldiers in battle (and, hence, friction in the military system as a whole) did, as a consequence of their military doctrine, penetrate to all elements of the military organism.

The result was a real awareness of the threat that friction posed to the functioning of the military machine. Under stress, a soldier's ability to perform his military function is reduced. In extreme cases, the Russians maintain, he will be completely unable to perform his military function, indeed to take any coordinated or purposeful action, even though he may not be physically injured.

Surprise, fear, fatigue, and pain are seen by the Russians as the main components of a stressful situation, a slightly different list than that drawn from Western studies, as indicated at the beginning of this chapter. When these components produce an excess of stress in a short space of time, the individual is viewed as being subject to shock, and his will to act is paralyzed. He will fail to perform his military function and become a prime cause of friction within the system. Moreover, if the system depends on him, the system itself will fail to function at that point. The effect of this friction on the army may be such that it is unable to take effective measures to counter the attacker's threat.

The Imposition of Friction on the Enemy

The production of sufficient friction to generate a "battlefield paralysis of the will" in the enemy is the prime goal of the Russian commander planning the operation. Accordingly, Soviet and Russian tactical and operational doctrine has always been governed by three main principles: speed of maneuver, concentration of effort, and surprise. The aim of the commander should not be to grind away the defender's strength but, rather, to deliver to his military system such a shock of surprise and weight of blow, followed by a rapid advance deep to the rear, that the defender is paralyzed by the resulting friction and thus unable to react. The defense, totally and suddenly undermined and destabilized at the very outset of war, will therefore collapse quickly even though a great many troops and units will still be individually viable. Shock, in other words, is used to incapacitate the enemy's military system by creating an overload of friction, so that the military machine is brought to a virtual halt.

To this end, the Russians hold, it is far better to achieve a surprise attack with minimum forces than to aim for a massive buildup that allows the enemy to steel himself for the onslaught, even though by taking the latter course, the attacker might ensure a greater degree of numerical superiority. The surprise and momentum will overwhelm the enemy staff system, generating such friction that it will be unable to cope with the volume of information it will receive in a very short space of time. It will consequently be unable to form a viable plan and produce an effective reaction. To compound this effect, the Russian doctrine insists that it is most important to deliver attacks simultaneously throughout the depths of the enemy position so as to increase the level of surprise and confusion and, hence, the instability of the whole defense.

This offensive against the defenders' rear, particularly against HQs, should create friction by three means: (1) physical attack, by bomb, shell, or sabotage squad; (2) electronic attack; and (3) rumor and psychological attack. Let's consider each of these in turn. The aim of physical attack is to generate friction by creating physical destruction of the command system so as to reduce the ability of the defender's military system to react; and fear amongst members of the staff for their own personal safety, thus increasing the burden of stress upon them. The aim of electronic attack is to compound this friction by preventing communication and thereby further hindering reaction; and creating frustration and a sense of impotence amongst members of the staff, further heightening the pressures on them and reducing their confidence and hence their resistance to stress. Finally, rumour and psychological attack complete the generation of friction, by being spread widely before the war, to sap the defenders' confidence in their ability to win; and during the war to heighten fear, mistrust, and uncertainty. Yet another Soviet and Russian principle of tactical and operational doctrine stresses the need to maintain *continued* pressure on the defender. This will deny the enemy time to rest and recover, and it will create extreme fatigue, which, as I noted above, is a most important feature in inducing friction in war.

As a practical way to achieve surprise and shock at the tactical level, the Soviets always stressed the use of new weapons and tactics. They figured that the enemy's resistance to stress would be weakened by encounters with the unexpected. Flame throwers and smoke screens were seen as tactical weapons to produce a high-shock effect. Chemical weapons were particularly valued because of the fear that even rumours of their use will inspire. Nuclear weapons, however, both of the high-blast type and of the enhanced radiation type, would cause by far the biggest shock to the system due to their extreme destructiveness and to the fear that the threat of their residual effect would create among the men. Note that even the fear of certain weapons would generate this friction, but it is important that the effects be rapidly cumulative, otherwise, the doctrine holds, the enemy would find a way to overcome the friction they produce. To use a new weapon or tactic on a limited scale, as the British did in World War I with the tank and the Germans did with gas, is to throw the advantage away.

In terms of planning the use of conventional weapons, this "shock" effect to the defender is considered just as important as, and in some cases even more important than, any physical destruction that use of a weapon or tactic might accomplish. Thus, in the realm of artillery fire planning, fire in support of the attack must be planned so as to achieve not so much maximum casualties as maximum shock. The heaviest barrage practicable in modern war would, the Russians say, incapacitate some 25 percent of defenders and equipment if these were properly entrenched. What matters, however, is not that the subsequent Soviet offensive would be met by fire only from the 75 percent of defenders still surviving; the number who ultimately die or who are hospitalized is virtually irrelevant. What matters is that for a certain time after the barrage ceases *none* of the defenders should be capable of firing their weapons at the attackers, including those not actually injured. The shock of the heavy shelling will paralyze and stupefy them, and the friction will have created sufficient inertia to stop the whole machine, if only temporarily, and the soldiers will still be sitting in a daze in their trenches when the first Soviet infantry arrive.[2] The Russians have always valued the multi-barreled rocket launcher for just this reason: The main advantage of its massive volume of fire (in the last war usually used to end a bombardment) is the *shock* it delivers to the enemy troops, rather than the lethality of its warhead.

Reducing the Effect of Friction on Russian Forces

The tactics of the Soviet Army and now Russian Army have been designed to give the system a high degree of resistance to stress and to reduce friction to an absolute minimum. These tactics are very well thought out and extremely simple, and far more details are reduced to basic drills than is the case in Western armies. The drills do not demand of the soldier a high degree of mental toil. If they are well learned, then to a certain extent they can be performed automatically. The first casualty of stress, the Russians maintain, will be clear and reasoned thinking. The last thing to go from a soldier's mind will be well-rehearsed drills. In essence, drill and repetition score over intellect, wit and initiative.

This realization certainly militates against the deployment of unnecessarily complex weaponry. If advanced technology is used, the Russians insist that it must serve to *simplify* the soldier's task and not to make it more complicated and therefore more a cause of friction. In one of the areas of highest technology on the battlefield—the jet plane—the application of this policy is very clear. First, Russian pilots have much less scope for initiative and are directed by ground or air-based controllers to their targets to a much greater extent than is the case in NATO air forces. Second, pilot training aims at the creation of a "dynamic stereotype in the cerebral cortex"; in other words, the training is geared toward producing complex reflex actions that do not depend on intellectual effort but, rather, are performed semi-automatically in the high-stress situation of the cockpit of a plane under attack.

A similar approach is taken toward the training of ground forces' commanders. A commander is taught that the offensive can be carried out in one of a limited number of ways. Consequently, when preparing for battle, he does not make a *plan* but, in Soviet terms, he makes a *decision*; that is, he decides which of his prelearned alternatives to employ, then amends his choice to suit particular circumstances.

Now much more automatic, the commander's actions are therefore less demanding and more apt to be performed quickly and efficiently under the stress and friction of battle. This method of operation is made even more automatic if the commander is taught to lay out his operational problems as "flowcharts" or algorithms, to help him come to a quick decision. The Russians are far from confident that they have solved the problem of information handling, decisionmaking, and planning. However, they are tackling this problem as one of the most important problems of the battlefield. Coalition experience in the Gulf bears out this assessment.

Tactically, of course, the Russian Army, like any other army, tries to ensure that it will not be surprised. The Russians recognize that one of their greatest vulnerabilities is being surprised, and they take great pains to make very detailed plans—based on the conviction that it is only by achieving surprise themselves, and by exploiting it very quickly indeed, that the enemy can be kept off balance and prevented from delivering a counterblow (for which contingency plans have not been made). *Achieving* surprise is therefore one of the surest ways to avoid *being* surprised (i.e., to avoid friction). Likewise, the Russians maintain that *early* success contributes to confidence building and the creation of a strong morale that will reinforce the psyche and enable it to more easily withstand the friction generated by stressful situations. Thus they place a very high premium on achieving preemption in a military operation.

A good example of tactics tailored to reduce exposure to stress can be seen in the Soviet and Russian organization of the march. The Russians consider the following measures to be essential when moving troops over long distances into contact: *first*, they must be protected against surprise attack by a security screen maintained around the marching column at all times; *second*, the march must be broken by substantial rest periods and the men *must* be given a chance to sleep. Deprivation of sleep for long periods is, in the Russian view, one of the most serious factors contributing to friction in war.

Moral conditioning of the man is another important bulwark against stress. The Russian language has no word for *privacy*. Thus it is possible that the loss of privacy, so inevitable in war, has less effect on Russians than on some of their Western counterparts, and is not the great creator of friction that it can be in Western armies. Morale in the Soviet and Russian armies has always been collectively generated and best maintained by group interaction. Indeed, the sense of the collective is very strong in Russia; it is deeply rooted in the traditions of orthodox Christianity. Loyalty to the collective therefore comes naturally to Russian soldiers, and the creation of teams is seen as essential to their subunit cohesion on the battlefield as it is in all armies. This extra stress on cohesion as a means of avoiding friction results in the employment of tactics intended to improve cohesive action and avoid subunit isolation. Street fighting and warfare in broken and hilly countrysides are seen by the Russians as very great threats to the viability of units because they force the deployment of minor subunits in isolation— with which the Russian mentality is, they say, ill-equipped to cope. Concern for this sensitivity to isolation is shown by recent restoration of the need for infantrymen in defense to dig two-man instead of individual foxholes for mutual *psychological* comfort and support. It is not just the Russian soldier who relies on the support of his fellows for his ability to continue to fight, but this support is perhaps more important for him than for his Western counterpart.

One problem that the Russians feel they face more acutely than some other armies is that they are, by nature, emotionally very volatile and prone to go from elation to despair very quickly. Resistance to stress, as a function of morale, can alter very quickly indeed, thus creating a high level of unpredictability in terms of how well the soldier will cope with, say, a reverse on the battlefield. During the 1970s, the Soviets invested considerable effort in the study of biorythyms in an attempt to predict individual cycles of depression and elation, and thereby to minimize the friction these would cause. The aim was to make the individual more aware of the changing nature of his moods and therefore more resilient because he would recognize these changes of mood for what they are. This interest in boosting soldiers' moods has fizzled out, but it stands as an example of the extent to which the Russians will go to reduce the unpredictability of battle—a factor that Clausewitz identified as a significant cause of friction.

Preparing the soldier for the horrors of the battlefield—the sight of death, the effect of fear on his own body, and particularly the noise—is seen by the Russians as being an essential part of psychological hardening. Without a real effort to prepare him for what he will experience, Soviet authors insist, the soldier inflicted with the shock of battle will not wish to go on. "Frequently in the last war," said the then head of Soviet combat training, General Salmanov, in an article in the late 1970s, "the attack halted because at critical moments of the battle attempts to overcome fear amongst the men failed." Certainly, in the last war, one of the most effective means of overcoming this fear and of forcing the Soviet soldier on was a draconian system of discipline. The Russian is by nature amenable to coercion. Force has always been a very effective means of rule in the country, and it has always been the single most important element in the army's disciplinary system. I could site many examples of officers keeping their men in line by brute force. In the Red Army during World War II, this action was legalized and formalized. Execution of a soldier on the battlefield was a matter of company punishment. It may surprise some readers that the application of force and the resulting creation of fear can be used effectively to overcome friction caused by fear from other causes; yet that is what the Russians describe. Of course, their disciplinary system does not rely on such means alone. There is an intense effort to inculcate in the soldier a fierce *hatred* of the enemy and a belief in the threat that the enemy presents. Backing this up, however, is a heritage of rule by coercion and a history of wars fought by means of coercion applied on the soldier, of which the typical Russian is fully aware.

Maintenance of draconian discipline in peacetime is much more difficult. Nevertheless, recent revelations of institutionalized bullying in the Soviet Army and a large number of suicides and exercise fatalities indicate that the army did maintain pressure on its conscripts. Today, such reports are seen by the population of the former U.S.S.R. as wholly negative and barbaric, and attempts are afoot to instigate far-reaching reform in the Russian Army. But humanity and decency put aside (as they often are in war), and from a purely military point of view, brutal training does not necessarily make for bad soldiers unless it is so brutal that it breaks the team spirit.

The foregoing has not been a comprehensive review of the Soviet and Russian military systems. There is much more to say about the role of the political officer and the KGB in creating as well as overcoming friction in war. However, the points made thus far are worthy of consideration in terms of our own military systems. There is, of course, no one single feature of the Russian attitude toward the stress of battle and its effect on man and the military machine that can be called revolutionary, or even one that is unique to the Soviet and Russian Army. What makes for a special Russian approach is a combination of attitudes; recognition of the problem and conscious effort to deal with it as a concept. The philosophical conclusion drawn from this realization that most distinguishes the Russian approach from the Western is that, although both recognize man as the most important element in battle and war, Western armies tend to rely most on the man as an individual. That is, they attempt to improve individual performance and to depend heavily on individual ability for effective unit performance so as to maximize the impact the man has on the battle. Russian logic dictates the opposite. According to this logic, if man is the most important element in battle and war, then as little reliance as possible must be placed on the individual, so that his loss or failure has only a minimal impact on the military system as a whole and, hence, on the outcome of the conflict. The military system must constantly be perfected to allow for this to happen. If war is likened to a game of chess, then the Russian would say that the Western trend is to concentrate on improving the quality, versatility, and initiative of each chess piece, whereas the Russian approach is to accept the limitations of the pieces on the board and put more attention into training the grand master to perfect his game.

One further quality of the Russian General Staff that we might learn from this is that, for all its imperfections and narrow-mindedness, it has always strived as a matter of principle never to forget the lessons of war and never to accept tailoring the army's performance to conform to peacetime political requirements. For this, the Russians must receive their due "which no man can imagine who has not seen war." In this short phrase, Clausewitz summed up the one point that, when we are discussing *friction*, differentiates war from all other phenomena and will condemn armies that forget this point to failure against anything but a grossly inferior enemy. It is not just that war is the prime example of friction at work, or that it is a more important social activity than any other. No, up to now we have been considering war as an abstract, as a social phenomenon in which the participants are "soldiers" or "officers" or "combatants"; we have talked of "systems" and technologies and strategies. But what makes war different from any other human endeavor, and what makes friction such an inevitable constraint of war, is that in war someone is trying to kill *YOU*. Please read those last eight words again, and ponder their meaning for a moment. This is not some abstract theory; it is not likely to be the result of any action by you; nor is it something impersonal. Rather, it is very personal: People you do not know, probably cannot see, and against whom you may have no personal contention will nevertheless be doing their utmost to kill, maim, or disgrace you, no matter what your rank or job. You will die and your family will be bereaved. Your death may pass unnoticed, unrecorded; you may be disfigured, disabled, or rendered insane; people may revile you and blame you, alive or dead. This will happen to you unless *you* and your colleagues strive to prevent it. This is what gives war its special meaning, and friction in war its special power.

Armies that forget this point will fail the test of war. Governments that find it politically awkward to maintain armies that make this a tenet of their doctrine neglect their duty to their people. Friction to armies is not only inevitable; it is also essential, like competition to business and, as I noted earlier, natural selection to the animal world. It is painful, uncomfortable; but in seeking to overcome it, armies succeed, whereas its absence leads to complacency, inefficiency, and ultimately failure.

References

Clausewitz, Carl von. *On War*. Translated by Sir Michael Howard and Peter Paret. Princeton, N.J., 1976.
Dixon, Norman. *On the Psychology of Military Incompetence*. London: Cape, 1976
Holmes, Richard. *Firing Line*. Harmondsworth: Penguin, 1987.
Keegan, John. *The Face of Battle*. London: Pimlico,1991.
Vigor, Peter. *The Soviet View of War, Peace and Neutrality*. London: Routledge, 1975.

PART THREE

Incentives for Progress

3

Let Us Now Praise Dragging Feet!

Ottar Brox

Sometimes those who try to lead, govern, develop or plan our societies do a good job because we are able to stop, delay, or pervert their schemes. More often we produce unanticipated problems and slowly developing catastrophes because we are too clever at eliminating sand in the machinery, such as "red tape" or "traditional attitudes" or "petty vested interests" of others.

Superficially, this chapter may be read as a conservative argument against ambitious attempts at comprehensive planning, or as a contribution to laissez-faire ideology. On another level, it may be read as a prescription for "more government intervention" into the civil society, advocating use of public resources to "arm the people", that is, to increase the ability of the little man to block the way of big business as well as big government. I perceive my position as a kind of amalgam among laissez-faire, populism, and "statism": The ordinary citizen is perfectly able to take care of herself, better able, in fact, than those protection agencies established to do so. As Murray Edelman (1964, ch. 3) and others have told us, such agencies tend to become servants of the interests they were supposed to control. Hence, it is more efficient to increase the citizen's anarchic ability to make life difficult for those who are out to change his world. But as we will see, this is a strategy that presupposes a government willing to intervene, vulnerable to parliamentary control, and thus responsive to the welfare needs of all citizens.

Let me illustrate these points by a reinterpretation of the history of Norwegian industrialization. Many books could be filled with the laments of Norwegian industrialists frustrated by the negative attitudes toward industry demonstrated by politicians, academics, artists, and the general public. Such laments would refer to strongly romantic or pastoral tendencies in the national literature, like the Nobel Laureate Knut Hamsun's ecstatic praise of the tillers of the soil or his aristocratic skepticism of the self-made entrepreneur or "nouveaux riche." Industry is generally associated with smog, greed, rootlessness, and moral decline. There are few industrial heroes in Norwegian fiction, but many heroic defenders of old virtues and ways of life. Lately, these tendencies have been reinforced by defenders of the natural environment.

This type of antimodernist skepticism has been expressed not only in literary terms but also in political oratory. In the words of Christian Michelsen (1857–1925), the Norwegian prime minister at the time of Norway's secession from Sweden in 1905:

> It seems as if the old rural economy is giving way to an emergent big industry. I have to admit that I consider this movement with great skepticism. If . . . no great industrial development can be carried through without [the emergence] of a large proletariat, I would consider it as a disaster to our country. . . . I think that a small people like ours would do most wisely to examine whether [a forced industrial development] is beneficial, and if it is not, if it can be hindered or obstructed. (1907; quoted in Wyller, 1975, pp. 157–158)

This "negative attitude toward modern technology" did not only manifest itself verbally, in political rhetoric and novels praising country life. More significant, the material development of the Norwegian economy was heavily influenced by what could be called active *Luddism*, as when smallboat fishermen physically destroyed a whale factory at Mehamn, Finnmark, in 1903.

The potential of *anarchic* Luddism to develop into *institutional* Luddism can be demonstrated by the "Battle of Trollfjord" of 1890 and its legislative repercussions. Since the Middle Ages, the entire adult male population of Arctic Norway used to gather in the Lofoten Islands in February and March to harvest spawning cod. Tens of thousands of small farm households got most of their cash income from participation in this fishery, carried out in open boats and with primitive gear. But the rich cod resources represented an opportunity not only to the fast-growing peasant population of northern Norway but also to the embryonic entrepreneur class: A shipowner sent steamships to the spawning grounds, closing off a whole fjord full of cod, which he then claimed to be the "owner" of. The thousands of smallboat fishermen did not accept this, and in a celebrated but almost bloodless naval show of force, involving hot steam hoses on one side and oars, fishgaffs, and boathooks on the other, the intruders were chased away.

At this time many of the fishermen involved already had the right to vote, and by the turn of the century all of them were voters. Keeping "outsiders" (e.g. shipowners, business interests, and trawler companies) out of fishing became a hot political theme, resulting in "populist" legislation preserving the fish resources for the peasantry for another three quarters of a century. Thus, spontaneous populist rebellions directed against modern technology were transformed into institutions that delayed technological and organizational modernization of Norway's most important export industry for generations.

The conventional perception of the strong antimodernist trends in Norwegian culture, as well as the struggle of the peasant population against the intrusion of modern technology, generally runs in this direction: These forces have been strong enough to *delay* the development of Norway into a modern industrial nation. The present leader of the Federation of Norwegian Industries has also blamed the anti-industrial forces for keeping the country dependent upon the export of primary products and resource-based manufacturing. Some spokesmen of the Federation have even found the century-old Michelsen tradition guilty of the deindustrialization of Norway after 1975.

Professional historians tend to be more cautious. For example, Bergh and associates (1983) cannot find any data to support the hypothesis that Norway's economic development has been delayed or hindered by the dominant industry-skepticism in Norwegian culture. Their overall view, however, is that modernization implies economic growth, which in turn has provided enough resources to avoid the predicted ugly consequences of industrialization, fast urbanization, and so on. In other words, thanks to the wealth created by modernization, the ugly consequences anticipated by skeptics like Michelsen have been avoided.

My thesis, in contrast, is that many of the divergent phenomena we refer to as "antimodernism," especially peasant resistance, have made important positive contributions to technological modernization and economic growth. By slowing down, the process has become more controllable and smooth, avoiding legitimacy crises and the political hazards produced by disregarding the fate of large *voting* elements of the population.

The deindustrialization of Norway during the last few years, even if partly explicable by the windfalls from oil, *can also be ascribed to the decline of the "antimodernist" forces.* Many of the more specific "progressive" effects of successful peasant resistance are easy to demonstrate. Let us reconsider the Battle of Trollfjord.

Harvesting the enormous shoals of spawning cod in the coastal commons was of course an attractive alternative for ambitious Norwegian entrepreneurs. There was a resource rent to cash in, as well as the revenue needed to cover labor and capital costs. If they had not collided frontally with the peasants—or, rather, if these peasants had been subdued by means of state power—profits to the shipping companies could have been enormous. But this attractive investment alternative was blocked by peasant resistance, anarchic and institutional, resulting in slower growth in the involved companies. The result was not retardation of the Norwegian economic development but, rather, the contrary. The capital of the Kaarbø family, which was kept out of harvesting natural resources by boathook-waving peasants, was redirected toward a slightly less profitable alternative. A shipbuilding business was established in the northern small town of Harstad.

The progressive aspects of investors having to stay away from easy and profitable harvesting of natural resources should not be too difficult to see. Getting the fish out of the water was a job that the peasants were perfectly able to do efficiently with their cheap and primitive small-scale technology. Thus, the capital accumulated by capitalist entrepreneurs was channeled to new, import-substituting branches of industry—such as shipbuilding—rather than tied up in primary harvesting. Without the "sand in the machinery" represented by successful popular resistance to modernism (i.e., steampowered fishing ships), northern Norway might have degenerated into a "resource region" such as a monocultural colony. We do not need hypothetical scenarios to see alternative routes of development. The world abounds in naturally rich regions made permanently underdeveloped through the easy access of capital into profitable resource harvesting—leaving the population of the resource region hanging in the air. The regressionary effects of industrialization on this level are most obvious when it involves undermining the livelihood of the regional population, as is obviously the case when capitalist fishing companies take resources away from peasant households.

If the peasants are able to take care of their "petty interests" and transform their "traditional attitudes" into populist legislation, accumulated capital may be canalized into new ventures, broadening the base of the national economy. But what about *new* resources, or resources not utilized in traditional peasant economies? Forests are a case in point. Sometimes, when new areas are connected to the world market through railways or waterways, large exploitation companies move in with their machinery and mobile labor, exploit the timber resources, and disappear, hardly affecting the regional economy. If this form of rapid economic development is successfully resisted by the population, the new opportunities may be used, gradually and slowly, to create nonagricultural opportunities for a scattered population too dependent upon a stagnant market for milk and grains. The friction represented by regional "traditionalism" may actually generate a more frictionless and progressive exodus of labor and capital from agriculture, whereas the population would be more tied up in stagnant branches of the economy if it had not been possible to put sand into the machinery of industrial modernism.

In contrast to forests, which can be exploited by small-scale technology—thus representing an opportunity to farm households—waterfalls can be harnessed only by economic organizations on a much larger scale. When hydroelectrical technology was sufficiently developed to make large-scale generation of electricity feasible—around the early 1900s—a prolonged struggle between "traditionalist" and "modernist" forces in Norwegian politics started.

"Modernists" eased the way into Norwegian fjords and valleys for large international companies, letting them buy waterfall rights from farmers at the local price level, hoping that this arrangement would speed up industrialization. The "traditionalist" forces prevailed, however, and concession laws were passed that limited the opportunities for industrial companies. First, ownership of the power plants had to return to the Norwegian state after a limited period. Second, a certain part of the resource rent had to go to the surrounding rural district. And, third, concession was to be given only if certain demands of Norwegian authorities (regarding location, employment, environmental concerns, etc.) were met.

The concession laws *may* have slowed down the growth of the electrochemical/metallurgical industry, even if a case can be made that these laws, with their overwhelming popular legitimacy, made industrial investment in Norway *politically safe*, even if less profitable than it could have been if the "modernists" had won. More important, however, are the probable repercussions of the revenue to the rural districts surrounding the large industrial installations. This money was partly spent to help small-scale primary industries, simple manufacturing, and job creation generally, besides communications, health, education, and other local services. Thus, larger industrial companies, whether Norwegian or foreign, did not became "cathedrals in the desert," offering wages far above the regional level or wages and other conditions of employment determined by surrounding rural squalor. Economic dualism (Brox,1972) was avoided, as was industrial development dependent upon abundance of labor at poverty level. When local conditions were favorable, labor went fast and smoothly from agriculture to small industries.

The aforementioned processes may be partly responsible for the seemingly anachronistic settlement pattern of Norway, compared with most other countries at the same level of industrialization. As late as 1960, when less than 10 percent of the British population and 25 percent of the Swedish population lived outside cities, *half* of the Norwegian population could still be termed rural. This difference is difficult to explain without reference to the fact that many categories of "traditional" adaptations had been defended by many means, from fishgaffs to concession laws. There are numerous cases in which large industries as well as governmental planning agencies have been frustrated in their efforts, and where we can demonstrate *post hoc* that the inertia resulting from successful resistance has speeded up real economic and social development.

A predictable argument against this analysis may be the following: Would not the industrial and economic growth have been faster, and less wasteful in terms of maintenance of an anachronistic settlement structure and so on, if modern entrepreneurs had been allowed to take over the harvesting of natural resources, creating a mass of available, mobile labor in the process? Up to the 1980s, the very bottleneck factor limiting the industrial growth process was considered to be availability of mobile labour, according to the employers as well as to many economists.

This is all very well, if we disregard the problem of political consensus. The peasants would simply not have *voted* for a government that had chosen a strategy in this direction. The dominant political force in the last century has been the Social Democrats and their rural-based liberal forerunners, and the rural vote has been necessary to any government alternative up to the present day.

The historical cases noted here may serve as examples of successful resistance to capitalist penetration of traditional economics, creating inertia that helped to channel private initiative into constructive directions. Modern social democracy emerged through these historical processes. The first three Labour members of Parliament were elected from Arctic Norway in 1903, more or less sent to the capital by their comrades-in-arms during the Trollfjord and Mehamn battles. The Labour Party helped to organize and institutionalize popular resistance, and its growth into parliamentary majority was stimulated through grassroot-level politization against capitalist growth as well as through trade unionism in the new industries. But ironically, the Labour ministers had hardly moved into their government offices when they met with the same kind of popular resistance that frustrated private enterprise under earlier, bourgeois regimes.

The Labour Party took over the government in 1935 and has stayed on since, interrupted for longer periods only by the German occupation and the Borten (1965–1971) and Willoch (1981–1986) cabinets. Although the party grew out of resistance to forced industrialization movements, its trademark has been *planning*, which first and foremost has implied speeding up the industrialization process.

A very characteristic case is the introduction of "industrial parks" in the 1960s. The Labour government, supported by most of the bourgeois opposition, had launched ambitious plans to develop "growth centers" in order to control and steer urbanization. Some were located in the central part of the country, designed to stop rural-urban migrants from ending up in the capital, whereas others were located in the provinces so as to maintain a certain national demographic balance.

The problem, as it turned out, was that naming a place a "growth center" and financing certain spectacular infrastructure projects did not set off any urban growth processes. New manufacturing industries preferred to locate in older industrial cities or in the home villages of small entrepreneurs, where conditions were more predictable. Hence, the idea of "industrial parks" was imported from Britain, and a state corporation was established to provide sites and buildings for rent or sale to private businesses.

There can be do doubt that these industrial parks were meant to speed up and facilitate demographic concentration. Keep in mind that, as there was no unemployment to speak of in the designated centers, the implication was that every new job had to be filled by people from other places. The intention was made even clearer by the condition, laid down in the original terms of reference, that the industrial parks should each accomodate 1,500 jobs at a minimum.

The implementation of this policy may serve as an instructive example of what Robert Merton called "dislocation of goals." Political actors pursuing very different goals were able to modify the whole idea of "industrial parks" to the extent that the effect was maintenance of the previous settlement pattern rather than forced concentration. Most important was the concession made to create a consensus about the industrial park policy: Such "parks" would be eligible for government support even if they could provide only a few dozen jobs. This meant that government money could be used to provide supplementary industrial employment in sparsely populated rural areas, reinforcing the previous settlement pattern rather than destroying it. The complete perversion of the "industrial park" concept facilitated the transfer of labor from primary to secondary industries, inasmuch as people could shift occupations without leaving their homes and conservative forces were able to sabotage government planning.

The development of Norwegian industries in the 1970s seems to prove this point. As Skonhoft (1982) has shown, industrial growth took place in a decentralized fashion, and certain forms of smaller industries in the periphery competed quite well with more centrally located units. The labor made redundant in stagnant primary industries was transferred to industry much more quickly than would have been the case if the workers had had to move. In short, successful sabotage of ambitious growth-speeding plans resulted in faster and less expensive modernization.

This success story can be explained by the ability of conservative rural "footdraggers" to dislocate the governmental planners' goals. But the story can also be told in a very different mode. In *Ulysses and the Sirens,* Jon Elster (1979) spells out one of the problems of government-by-consensus—namely, that open political systems tend to make tough but necessary decisions difficult to carry through, and planning becomes illusory. A political body may accept a plan, but too many actors, even if they were represented in the political body, may, by their aggregate efforts, nullify or completely pervert the plan in the process of its implementation. This mechanism can be observed at all levels, from national wage policies to local building regulations. Let's say a municipal board accepts a zone plan that prohibits the building of dwelling houses close to the highway. But when A applies for a building permit, his relatives, co-religionists, football friends, and party chums are able to form a majority coalition that makes "an exception," and the same happens in the case of B, C, and D as well. After a few years, the road is full of playing children (Aaseth, 1980).To make democracy efficient, authorities must be rendered immune to perverting influences after the course has been set, just as Ulysses had himself tied to the mast and his ears waxed to avoid ending up on the rocks.

In these terms, the story about the Norwegian import of the "industrial park" concept would be a story of failure, not success. The assumed "unavoidable restructuring of the national economy and settlement structure" was postponed, and the preindustrial distribution of people and installations reinforced and maintained, at the expense of urban growth. To maintain consensus among the crew, the helmsman has to listen to the sirens, straggle from his course, and end up on the rocks.

The "Ulysses" analogy would have been perfect if we always could be certain that the man at the helm was right, and that all suggestions raised during the passage would lead the ship in wrong and dangerous directions. But sometimes catastrophes, calamities, and wasteful schemes are avoided because the ears of the helmsman are free from wax as when the political system is open and flexible.

A "Ulysses" interpretation of the story assumes that the helmsman knows what in the long run is best for the crew, and that the sailors have anticipated and evaluated all implications of the navigation plot. But in politics, at any rate, the "Ulysses" analogy tends to mislead. Very seldom do the alternatives to following a government plan result in total national disaster. Most real-life "Ulysses" stories can be told in terms of plans being accepted without considering alternatives, and by decisionmakers (e.g., Parliament members) without sufficient time, professional assistance, and imagination to anticipate even the predictable implications of the plan. Ambitious goal-steering is based upon illusions about human abilities to anticipate and predict; it is also difficult to reconcile with real democracy, which presupposes the ability of the common woman and man to stop, delay, or pervert the schemes of the rulers.

And *nota bene*: There is not necessarily a trade-off between democracy and efficiency, in the sense that one might choose opportunities for democratic intervention rather than faster growth and a higher standard of living. I hope that the cases related in this chapter demonstrate that successful popular resistance and successful attempts at dislocating government goals may speed up real growth, for example, through easier transfer of factor inputs from stagnant primary industries to more promising secondary ones.

I have not told the whole story, however. Blocking the access of big capital to rural resources and perverting the overambitious plans of big government may be necessary preconditions for sound and smooth development. Obviously, grassroots power and resistance ability are not always sufficient preconditions for growth. But in the Norwegian case, the organizational abilities developed during the various resistance movements were put to use in other arenas as well. Small primary producers not only stopped the invaders; they also organized cooperative marketing and other arrangements to take advantage of new technology and economies of scale. Moreover, the rural and small-scale producers' parliamentary influence was used to press for extension services and financing schemes to "industrialize the countryside." Even if planners in high places did not believe in it, concessions to "alternative development" were necessary in order to maintain parliamentary majorities.

As I interpret the history of Norwegian industrialization, it is the "friction" or "inertia" provided by popular, anarchic, and institutional resistance to modernism that can explain its relatively smooth run and avoidance of negative social implications. Indeed, it is tempting to explain the deindustrialization of Norway since the 1970s with reference to the same variable. The oil flow from the North Sea has eliminated the friction referred to earlier, thus destroying the delicate system of checks and balances that has characterized Norway's development into an efficient, modern, industrial welfare state. That part of the story must be told elsewhere, but I will conclude with a few illustrative examples.

Small rural fishermen will generally fight to keep their right to fish, but maybe not if there are good welfare alternatives. Generous relief and unemployment benefit schemes, financed by oil revenue, make it easier for large trawlers to take over the marine resources. The friction retarding "modernism" has thus been removed by oil lubrication. Small-scale decentralized milk production, utilizing production factors with very low shadow prices, is easily taken over by heavily subsidized, centrally located "agribusiness" when capital subsidies to the new farmers as well as different forms of public compensation to the old ones are easily obtained from oil funds. Small funds, set up to finance rural industrialization, are easily integrated into national big-industry finance institutions when rural politicians are given much more generous funds to finance welfare, schools, and other infrastructure. Reduced friction also means that the access of big capital to resource-based branches (not only oil but hydropower, fishing, and aquaculture as well) has become easier, thereby reducing the investments in competition-exposed manufacturing industries. This was especially the case during the first phase of the oil age, when the country developed a rent-based, speculative economy, rendering Norway rather vulnerable to adverse business trends.

The functions of friction in economic systems are badly recorded, insufficiently understood, and barely included in economic theory. Those who believe in the magic of the market seem to be no less blind to the benefits of sand in the machinery than those who believe in ambitious government planning. The reason seems to be a familiar one: What is good for the whole society, in the long run, is not always good for short-sighted and influential actors with power to control the definition of the situation.

The consequences of our insufficient understanding of the benefits of friction are all too obvious in many Western countries in the 1990s. Urged on by the best-organized vested interests, governments struggle to invent ways to reduce friction. In Norway's case, at any rate, this is like killing the hen that used to lay golden eggs.

References

Aaseth, K. M. "Planlegging, kommunalpolitikk og lokal organisasjon". *Tidsskrift for samfunnsforskning*, vol. 21, 325–335. Oslo, 1980.

Bergh, T., et al. *Norge fra U-land til I-land*. Oslo: Gyldendal, 1983.

Brox, O. *Newfoundland Fishermen in the Age of Industry: A Sociology of Economic Dualism*. Canada: ISER St. John's,1972.

Edelman, M. *The Symbolic Uses of Politics*. Urbana: University of Illinois Press, 1964.

Elster, J. *Ulysses and the Sirens*. Cambridge, England, 1979.

Skonhoft, A. *Industriens regionale omstrukturering*. SINTEF-rapport. STF83-A82011. Trondheim: NTH, 1982.

Wyller, T. C. *Christian Michelsen. Politikeren*. Oslo: Dreyer, 1975.

4

Social Change Induced by Technology: Promotion and Resistance

T. R. Lakshmanan

Technological progress, offering an increase in output that is not commensurate with the increase in the costs necessary to generate it, takes the form of that rare thing in the economist's lexicon—a "free lunch." Countries or regions that take advantage of this "free lunch," through sustained technological progress over time, ratchet themselves up to high levels of prosperity and modernize their social and economic structures. Notable examples of such sociotechnical transformation can be found in Britain since 1775, the United States since the early nineteenth century, and Japan and Sweden in the last century or so.

Stimulated by these and other successful cases of development, many developing countries have attempted in the last four decades a transfer of modern technology in order to lift themselves out of poverty into self-sustaining development. Some newly industrializing countries (e.g., Korea, Taiwan) have achieved moderate to high levels of technological catch-up with the advanced industrial countries and rapid rates of real income growth, whereas many other developing countries are lagging behind and failing to develop. Accordingly, there is considerable intellectual and policy interest in the experience of late industrializing countries that have engaged in successful experiments in technology-based development, in order to identify the conditions under which dynamic self sustaining development takes place.

Theories of Development

Self-sustaining development and the massive social transformation attendant upon nineteenth century technology-led industrialization have inspired a variety of theories of development and the attendant transformation known as *modernization*. These theories derive from many disciplinary sources—largely from sociology and economics, but also from political science, anthropology, and business psychology. Some originate in the liberal (and neoliberal) bourgeois framework, others in Marxist (and neo-Marxist) versions of the European and North Atlantic development experience, and still others in the more recent transformation experience of the peripheral developing societies in the international system. From this extensive literature I have culled some major ideas that have evolved toward a contemporary view of development and modernization.

Economic theorists, as noted earlier, have traditionally emphasized growth (which refers to an increase in the quantitative level of activities and the scale of its social structures) and *optimizing behavior* in a given social and technical context. Since the essential character of technical change is that it leads to change in the mode of behavior of an activity system, there has been a recent concern with development (rather than growth) as defined in a variety of ways: in terms of an augmentation of the concept of growth to include greater capacity for generating consumption goods and their more equitable distribution (Chenery and Ahluwalia, 1974); in terms of the meeting of basic needs such as food, clothing, education, housing, and health care (Streeten and Associates, 1981); or in terms of levels of individual functioning and capabilities (Sen 1988).

For our purposes, however, the concept of development is not an end state but a *process*. Development implies a dual structural change: (1) the emergence of a new social and technical environment or a new set of economic opportunities and (2) a changed pattern of behavioral relationships between the environment and the social actors. Viewed this way, development induced by technical change encompasses the following changes in the social and technical environment: the technical transformation of agriculture, industrialization, and urbanization from the space economy perspective; the weakening of kinship ties and the rise of achievement orientation on the social side; the rationalization of authority and the growth of bureaucracy (which is both enabling and constraining for the individual) on the political side; and, culturally, the rise of science and secularization in society resulting from increasing literacy and numeracy. Development and modernization thus represent a transformation of the material, spatial, structural, and cultural conditions of society. So we need theories concerning the processes of development and modernization.

The first is a theory of endogenous social transformation based upon the experience of development in nineteenth-century Europe and North America. In this model, modernization was viewed as endogenous in the sense that society was deemed capable of transforming itself from within, but not because of intervention or pressure from without. Culture (in the sense of knowledge or a set of *situational models* of behavior and thought) changes first. Then new forces of production and techniques emerge — specifically, through increasing secularization, differentiation of institutions, and greater economic individualism. Finally a mobilization of physical, human, and organizational capital occurs, all leading to social transformation. This model was subsequently elaborated as a general model of modernization toward which all countries would progress at different speeds from their different points of departure.

On closer examination, this idea of endogenous development could not be sustained to describe the experience of late industrializers. In Gerschenkron's (1966) study of modernization of continental European countries before World War I, the role of the state was crucial in coordinating the late industrialization through technological borrowing and the creation of the " functional equivalent" of economic and social institutions necessary for development. (No doubt among the first national states — England and Sweden — the state was primarily the guarantor of civil peace, as well as of the free movement of ideas, people, and goods. For later national states, the state has become an "active" power, directing not only economic activity but also a growing number of social aspects of life.) The national state, whether run by a bourgeoisie or by a modernizing elite (Germany, Japan, Italy, etc.) or a socialist regime (the Soviet Union), remained the modernizing agent. A survey of the development experiences of Third World countries suggests that their modernization is even less endogenous. Whereas the modernizations initiated in both the "Western model" and the "socialist model" were essentially guided by indigenous national groups acting on behalf of the state, in most parts of the Third World modernization was exogenous, arriving with the influx of foreign capital and the material and nonmaterial infrastructures associated with the international system of trade and production.

Indeed, this realization — that the processes underlying technology-led transformation is dependent upon the manner in which the various developing countries have been "inserted" into the international structure of economic, political, and cultural relationships organized by industrialized core countries of Europe and North America — has spawned a second group of theoretical formulations: "underdevelopment," "dependent development," "unequal development," and so on. These formulations range all the way from Andre G. Frank's extreme view of the iron cage of core-periphery relationships leading only to "underdevelopment" of peripheral developing countries, to the far more sophisticated formulations of the differential development experiences as contingent outcomes of interactions between indigenous social groups and core groups in the context of international flows of technology and capital (Cardosa, 1972; De Palma, 1978). Such formulations, which developed in opposition to the earlier universalist view of development in the Third World, have in turn led to region- or culture-specific models of development. Two main points need to be made about these two broad streams of theorizing on development and modernization.

First, the endogenous theory of modernization described earlier emphasizes universalism, a derivative of the nineteenth-century notion of modern society with its capacity to apply "the universal principles of Reason to all particular situations." This idea of social evolutionism has been pushed further by social thinkers such as Marx (with emphasis on the "laws" of historical development capable of accounting for most aspects of social life and its transformation) and, in this century, functionalist scholars who characterize the situation of most countries by the obstacles that develop in accordance with the universal model of modernity, as expressed in advanced Western societies. This widely prevalent theory of convergent social evolution (supporting the notion of inevitable "convergence" of all societies being transformed by technology change), that has persisted from Marx in the last century to recent times, is an unproven premise and basically flawed (Dore, 1986). If that theory were true, the two societies that adopt the same cluster of technical innovations and get transformed in the process, should be associated with a similar type of institutions. This condition does not hold even for basic institutions such as those surrounding the organization of industrial work. Two societies transformed by modern technology—early nineteenth-century New England and late nineteenth-century Japan—provide examples of different motivational structures (institutions) and informational structures (incentive systems), i.e., quite different ways of recruiting, educating, motivating, managing, and rewarding industrial labor, while adopting the same set of mechanical technologies. The unproven premise of universality in these models is, as Hirschman (1971) notes, "a hindrance to understanding."

Second, the responses to the universalist view in the form of culture- or region-specific models of development in the subordinate peripheral countries are again an inadequate formulation. No historical change or passage from one type of society to another is purely endogenous in the sense of change occuring simply as a result of the simple accumulation of technologies, wealth, or various exchanges. Since even the most advanced societies are exposed to external technical, social, and cultural impulses, social change results from a combination of endogenous and exogenous sources. Similarly, in peripheral developing societies exogenous technical impulses from the larger world generate considerable changes, in the form of capital accumulation and new wage labor but also of new social interest groups, new rising expectations, and new contradictions. Those countries that "seize the chance" and successfully develop (Korea, Taiwan, etc.) take advantage of opportunities released by the forces of change in the international system and combine both universalistic and regional elements on the one hand and economic, socio-political, and cultural elements on the other. Korea and Taiwan, in particular, combined the opportunity posed by the relatively open access to U.S. markets and technology, an industrialization strategy consistent with their initial and changing relative factor prices, and the cultural attributes of the so-called East Asian development model (Berger, 1991).

If one adopts such a formulation of social transformation as a combination of generic and specific elements, it follows that there is one *modernity* that belongs to *many types of modernization*, depending upon the culture and type of society undergoing modernization (Touraine, 1988). Modernity refers to the capacity of a society to innovate, to create, to make relevant (economic and social) investments, and to engage in self-transformation, a capacity that is driven by technical and social knowledge. Modernization is the process of mobilizing societal resources for effecting technical and social transformation. It involves the matching of the characteristics of the new technology with the factor endowment and social structure of the existing society. The manner in which this matching process takes place can and does differ from region to region, each of which may exhibit different factor endowments and social structures. Regions differ in terms of being heirs to different historic, cultural, and institutional styles and engage in different processes of mobilization. There is clearly no universal value to the course of technical and social transformation of the first region (North Atlantic). Indeed, as I discuss later, societies outside the North Atlantic region have modernized successfully while utilizing somewhat different means—borrowing models of knowledge and structure but imbuing them with a blend of indigenous and imported meanings, and providing different forms of motivation and different support structures.

The elements of a new conception of social transition are available from recent work of such disparate sources as general system theorists (Vickers, 1973), development theorists (Dunn, 1970) and social theorists (Touraine, 1988, Giddens, 1979). Although these theoretical constructs focus on different aspects of the transition (system change, social learning, cultural change, social and ethical conflicts, etc.) and use different descriptive categories, they exhibit a remarkable convergence toward the idea that social transformation results from the social and cultural interactions among groups of social actors mediated by power. During the transition, some social actors draw attention to the growing gaps between the existing models of thought and behavior and the requirements of the changing economic and social conditions, and call for change. This protest alerts the unaware and mobilizes the defenders of the status quo. In such contests, successful agents of change use their new knowledge and resources to make a variety of economic and noneconomic investments in the interests of changing culture and creating new relevant social and economic structures in the context of active resistance to their ideas. I elaborate on this idea in the next section in order to delineate the nature of societal transition, the dynamic processes involved in the push for change, the frictions they encounter, and the contingent outcomes of such contests.

Social Transition and Resistance to Change

Our culture defines the way we view the world and the range of options open to us, including the actions that change our culture. The culture itself encompasses characteristic ways of making things and organizing action as well as a set of situational models of behavior and thought. Culture is not a once-and-for-all screen between us and the outside world but, rather, an ongoing process, continuously constructed and reconstructed during social interaction. There are two complementary processes at work, in the sense that each of us is being socialized into and differentiated from the culture. The first process provides coherence and meaning, and the second is a hedge against rigidity. When culture is deemed inadequate by significant subgroups in society, they press for change. They protest against the inadequacies, and what was once a tacit and unquestioned position becomes the focus of heightened consciousness and debate.

It is in this context that the notion of agency is relevant. Agency, in social theory, refers to social categories, actors, or classes who are capable of changing the relations of power and dominations and, hence, the prevailing arrangements in society. In Touraine's (1988) view, social actors view culture as a set of resources and models, a stake, that they attempt to manage, control, and appropriate, and whose incorporation into social organization they try to accomplish.

As new knowledge and new techniques emerge, changes in the forces of production and social mores result. New social actors associated with new knowledge and technologies come into being armed with radical ideologies (examples include the industrial entrepreneurs of the nineteenth century and intellectuals with a key role in knowledge production in the postindustrial society). Such social actors push for the rapid diffusion of the new technologies and highlight the discrepancies between the "functional requirements" of the emerging milieu and the existing models of action and thought. The latter models constitute sets of interrelated models—models of knowledge, investment models, and ethical models— that have developed as responses to social problemsolving of an earlier era, when the material conditions were quite different. Thus, during the transition to industrial society, for instance, new modes of knowledge, new normative orientations, new economic investments, and new social organizations appropriate to the functioning of an industrial society emerge and become increasingly relevant to considerable parts of society, while the dominant forms of cultural and institutional structure remain those of the preindustrial society.

It is in this context of a transformed culture and forms of social organization attached to the past that the social actors, in the economic, institutional, and cultural realms, who are spearheading industrial activity, push for revision of the older orientations and institutions. As active carriers of new ideas, they offer radical proposals in terms of new normative orientations, economic philosophies, and innovations in the organization of exchange, production, consumption, and political activities. Although there may be competition among these sets of social actors, they share a common resolve to change the existing system and a feeling of relative deprivation with regard to existing loci of power.

Such an articulation of protest, questioning, and advance of proposals for new social relations stirs up debate and criticism and educates the unaware. These outcomes, in turn, mobilize the defenders of the status quo. The resulting arguments between the agents of change and the custodians of the extant system are not civil "conversations." The conflicting ideas and allegiances are entrenched and defended: The groups in conflict have clear interests and the means to fight for them.

Thus the stage is set for mobilizing resistance to social change on the part of those groups who will suffer change and loss in the material and nonmaterial realms from the modes of accumulation and institutions being advocated by the agents of change. The resulting conflicts between the agents of change and the existing normative and institutional legitimators occur at several levels, with the situation remaining indeterminate and constantly changing. One would expect the technologies that do not disrupt existing institutions (i.e., conservative technologies) to be accepted. But those technologies that, in contrast, alter, transfer, or displace the prevalent institutions and orientations will be resisted.

It is not clear, before the fact, which of these contested models will prevail, what form they will take, or which groups will emerge to control these changes. But I can illustrate the nature and magnitude of these contests in terms of the kinds of changes that took place in the normative premises and institutional structure in the transition to industrial society. In that transition, with its central idea of transforming the conditions of production, a strong belief in work, energy, deferred gratification, savings, and progress was necessary. As Max Weber pointed out, the shift to these values took place in the North Atlantic Region (particularly in Calvinistic areas) via a religious radicalism *prior to industrialization.*[1] Consequently, in Western Europe and North America, much of the change in the normative models preparatory to industrial society had taken place *before* the arrival of the technologies we define as industrialization, thereby lowering the resistance to change.

In societies in other cultural realms that industrialized successfully later, such normative shifts from "traditional" to modern orientations had to be created, resulting in a higher level of resistance from holders of prior normative positions. Indeed, as I discuss later, this major change in the symbolic premises of the society was achieved in Japan by a resourceful modernizing elite, who invented the functional equivalent of the Calvinist will to work in the form of a drive for a strong Japan and a creative use of various symbols of the past.

In such change-oriented strategies, institutions become both the target and the *instrument of change.* It is the anomalies manifest between the goals of industrial society and the operation of social system controls (as specified by older institutions) that constitute the motive for targeting institutional change. The agents of change visualize social innovations (e.g., market innovations, managerial innovations, political innovations and institutional innovations) that would improve the consonance between goals and controls through experiments in social reorganization. This process of attempting to embody the new normative orientations in an effective way in the institutions was described by Dunn (1970) as evolutionary experimentation. In terms of our present subject matter, evolutionary experimentation takes place when social actors as change agents seek the progressive solutions of perceived social anomalies. In this context we have social learning, which embraces both individual learning and the learning that attends the transformation of social systems.

Institutional reorganization also becomes an instrument of social change as more and more individuals and groups come under the sway of the new institutions and adopt new modes of knowledge, the new ethical models, and the newly desired economic and social investment "practices"

Overcoming Frictions During the Social Transition

Promotional Strategies

Before we elaborate the notion of material and nonmaterial *infrastructures*, let us recall the frequent observation that the modern industrial system created in the last century in Europe and North America was not merely a technology of modern production (Rosenberg and Birdzill, 1986). The deeper function of a modern industrial enterprise was not simply to operate a factory efficiently but to create constantly those improvements (in product, process, organization, or inputs) that would increase profits. In other words, *enterprising* and *continuous innovation* is the key to this system for continuous change. To be a truly modern industrial country, a developing late starter must therefore not only transplant new technologies *but also erect a system for continuous change*.

If the essential characteristic of a modern industrial country is the social capacity to make continual changes in that country's physical and social technology, successful developers must acquire this capacity over time. I suggest that the acquisition of this social capacity during the transition from a preindustrial to an industrial society is possible only with *fundamental changes* in the *determinants of that social capacity*—namely, the material and nonmaterial infrastructures of the society.

We view infrastructures not in the traditional sense of artifacts or things *but as a set of attributes* (Lakshmanan, 1989). These attributes reside in some material or physical infrastructure capital (e.g., in networks of transportation or communications), and in some nonmaterial infrastructure capital and networks (e.g., in knowledge structures and networks, ideology, social mobility, rules for political and economic decisionmaking, etc.) So, what we view here as nonmaterial infrastructures are components of cultural systems and social and political institutions in a society. The relevant attributes of material and nonmaterial (hereafter called M-N) infrastructures are their (1) long duration, (2) collectivity or publicness, (3) central role as connectors among and between people and things, and (4) capacity to define what is possible (i.e., the set economic opportunities) at a given level of technology.

Thus, material infrastructure (railroads, telecommunication systems, electric grids, etc.) can be used for the movement of goods, ideas, energy, and goods and services. A non material infrastructure such as education is used for the production a wide range of goods and services, but also for evolving new ideas and new combinations of factors of production and for promoting innovation. Again, in a traditional society, other forms of nonmaterial infrastructure can promote greater social mobility and release latent energies for individual and social development.

My concept of infrastructure and its relation to economic and social performance therefore differs in two significant ways from traditional formulations First, I go beyond the common view of infrastructure as physical capital (e.g., transportation, communication, power, etc.) to include nonmaterial capital (e.g., knowledge structures and networks, social ideology, institutions, etc.). Second, I relate infrastructure thus defined not to growth (scale changes) but to development (structural changes) of the economy. In short, I argue from my analysis of case studies of successful developers that *a set of investments in physical, cultural and social infrastructures converge to create new resources, and new sets of opportunities for social thought and action, simultaneously facilitating and constraining actions of households, organizations, and the entire society.*

What are the functional capabilities sought and provided by material and nonmaterial infrastructures? These functional capabilities are necessary conditions for development or structural change, and in cases of transition to industrial society they include the following:

1. *Knowledge and material diffusion structures* such as the material infrastructures (canals, railroads, roads, etc.) that facilitate the movement of materials, goods, and people necessary for industrial production; or the system of education and support structures for training and acquisition and diffusion of new knowledge.
2. *Social mobility structures*: such as new fundamental (constitutional) rules for social interactions that facilitate the transition from inherited rigid, status-based structures to more fluid social structures permitting freer economic opportunities based on achievement.
3. *The passage to the market*, involving, for example, the reorientation of society from premodern, typically kinship-based social incentives (guiding individual and group actions) to more exchange-based incentives.
4. *Normative shifts*

Compensatory and Nurturing Strategies

This second set of strategies tackle social resistance head-on in two ways. First, they soften the economic and social costs of change by providing some economic and noneconomic compensation for asset and status losses. Second, they build bridges between old and new activities and institutions in such a way that the increasingly obsolete is not destroyed outright but, where possible, is modified and related to the new. In these two ways, the older institutions and activities can develop some stake in the radical transformation and attenuate the frictions in the path of change. In the next section is a discussion of my formulation of such social transitions, highlighting both the strategies that sponsor and promote change and the strategies that lower the resistance to social change (i.e., friction) as exemplified by the conversion of Japan into an industrial society in the last century.

The Case of the Japanese Transition to Industrial Society

Japan's transformation is well known. From its humble beginnings as a poor subsistence economy, characterized by feudal restrictions on labor movement, limited autonomy of economic sectors, and a lack of modern material and nonmaterial infrastructures, it developed into a great economic, military, and naval power in six decades (Beasley, 1990; Nakamura, 1981). Since the end of the World War II, which resulted in major material losses, technology-led economic growth resumed at an accelerated pace and transformed Japan into an economic superpower, accounting for 11 percent of the world's output.

The early modernizers of Japan built upon the legacy of the premodern Tokugawa era (1600-1868) as expressed in knowledge, culture, and ideas as well as economic and social strengths and obstacles. The Tokugawa period was a form of "national" feudalism centered on the shogun in Tokyo—the *Bakufu*—with a rigid, fourfold, Confucian-type hierarchy of samurai, peasants, artisans and merchants and a xenophobic closing of the country to outside contacts. Toward the end of the period, the long era of peace and the Tokugawa institutional structures created conditions conducive to major economic and social changes and, indeed, to formation of an "alternative society" parallel to the feudal structure (Lakshmanan, 1990). The institution of "alternate attendance" (Sankin Kotai), whereby the lords had to maintain residence at Tokyo in alternate years, had led over time to a workable road network, vigorous coastal shipping, an impressive urban network, and a standardization of language and custom—in other words, to a physical and cultural base for building a nation state in the Meiji period. In addition, the separation of the samurai from the land and his use as a civil official (a use that was increasingly impoverished) led not only to a large well-educated bureaucratic class but also to social fluidity and a disaffected warrior class. In 1868, the Tokugawa period left behind a society with some feudal constraints but also with a well-developed agriculture, remarkable rural industrial development (analogous to the U.K.'s proto-industrialization or to Germany's Verlag system), an extensive transportation system, respectable commercial infrastructure, and a well educated population with a strong work ethic and a range of industrial and bureaucratic skills.

The Meiji Modernizers: Composition and Orientations

A variety of interconnected political, social, and economic discontents (against a backdrop of Western economic penetration) surfaced in mid-nineteenth-century Japanese society, expressed at times in writing and at other times in action. The proposals for overcoming these discontents were not framed, as in Europe, in public debate about social justice or the desire for a new political system; rather, they were formulated in the familiar vernacular of the emperor versus the rule of the shogun. The agents of change who emerged from these internal struggles of the last years of the *Bakufu* derived their ideas from the countermodels offered by two intellectual strands, which, though somewhat in opposition to each other, tended to coalesce. One came from the exposure to Western economic, social, and political institutions brought about by the Dutch studies program (for gunnery and medicine). The relevant ideas of reform of Japan spread largely among the low- and middle-level samurai of the southwestern peripheral clans (*han*)—the *Satsuma, Choshu,* and *Tosa*—who had been most exposed to the foreign menace as well as to foreign influence in terms of Western science, technology, and weapons, and who had to adopt some modernization programs. In the latter process, low- and middle-level dissatisfied samurai had acquired economic and administrative experience and positions of political influence. The second intellectual strand came from the scholars of National studies who stressed nationalistic positions intended to resist foreign intrusion and to support the imperial institution.

For the modernizing group, Japanese political and economic independence could be preserved only by "wealth and strength"—"wealth" through adaptive and selective modernization of the economy and "strength" through Western weapons and military organization. This group (helped by other influential groups, including the leading merchants) seized power in the name of the emperor (during the Meiji Restoration) and exercised that power as bureaucrats acting through government machinery, which became increasingly Western in its orientation (Beasley, 1990). In the first decade of the Meiji era, these young samurai from the peripheral *han* who formed the higher bureaucracy shared power with the imperial court nobles. In short order they emerged as ministers and began investing heavily in what I have described as material and nonmaterial infrastructures, thereby fundamentally changing the economic, social, and political arena. I describe their resourceful strategies next.

Nonmaterial and Material Infrastructures for Promoting Social Change

I view the investments in these infrastructures by the Meiji modernizers as directed by the following objectives: (1) facilitating the passage to the market, (2) creating the ideological basis for modernization, (3) creating a progressive entrepreneurial class, (4) organizational restructuring, and (5) facilitating easy movement of people, ideas, and materials.

The Passage to the Market

Market exchange is the *dominant* transactional mode in a modern industrial economy; that is, it integrates the patterns of economic interactions among individuals and groups. Basic resources—labor, land, and capital—that enter into production are organized for sale in markets. Indeed, resource allocation, work organization, and product disposal are arranged through this mode. Accordingly, the inherited incentive structures (which guided individual and group behavior in the Tokugawa period and were based on the principles of reciprocity, friendship, kinship, status, and hierarchy) had to become more supportive of the economic transactional mode of market exchange. In the first decade of the Meiji Restoration, the abolition of the elaborately formal Tokugawa status society and the rigid class system in favor of an achievement-oriented society made for a more level playing field for every Japanese. Since feudal restrictions on the ability of a person to move from the area of domicile were also removed, along with restrictions on occupations and on doing business in guilds, an individual from the former peasant, merchant, or artisan class could, abilities permitting, take advantage of any economic opportunity that emerged during modernization. With the right to private ownership of property, the ability to buy and sell land freely, and land tax reform, farmers could invest more heavily in land and retain more of the increased productivity, thereby spurring accumulation. The army, the navy, the government bureaucracy, new commercial and industrial enterprises, and professional work were now open not only to former samurai but to members of other classes as well.

The promotion of self-improvement that was *individuated rather than individualistic* in Meiji Japan occured in the context of maintenance of interdependence and harmony in relationships. Thus competition and social mobility, so vital for economic progress, were encouraged among individuals who are viewed as parts of groups. *The Meiji liberation of the individual was thus in the Japanese idiom, entailing a unique blend of indigenous and imported meanings of the liberated individual.*

The Ideological Basis for the Passage to A Modern Technical Society

Since an economy is embedded in a society and its culture, more than pecuniary incentives are involved in the transition to a modern technical society. Scholars ranging from Max Weber to John Maynard Keynes have emphasized the pivotal role of belief systems in guiding modern economic behavior. In the Weberian model of the West, religious (Calvinistic) beliefs—vertical and nonmaterialistic values—which developed in the process of breakaway from the Catholic Church, engineered a break from past values and impelled the individual toward rational, regular, and constant work for profit and accumulations. How did Japan functionally substitute for the role of Calvinistic ethics in effecting a break with these past values and creating a normative structure favorable to rational, constant work and unlimited accumulations? As Michio Morishima (1982) suggests, the two main religions in Japan, Buddhism and Shintoism, have had little or no influence on secular or economic conduct. However, Confucianism, emphasized during the Tokugawa period, is an intellectual and rational ethic (rejecting the mysticism common to most religions). Although it discouraged individualism, its rationalistic perspective encouraged Japanese openness to Western science and technology. Whereas Chinese Confucianism emphasized benevolence, the Japanese variant of Confucianism highlighted loyalty in the individual's relationship to various communities—to state, to emperor, to family, and to friends (Morishima, 1982). It was this strong sense of loyalty, inculcated over the centuries, that was utilized in the shift to the new ideology instrumental for modernization. Engagement in modern business activities for the greater glory of Japan and the Calvinist drive to work for one's salvation and the kingdom of God were very comparable schemes for effecting the desired change in the normative structure directed toward economic achievement.

In order to spread the ideology of progress and development widely in a short time, the Japanese developed various slogans indicating full acceptance of Western technology and ideas. One such slogan was *bummei kaika* ("civilization and enlightenment"). Analogous to Japan's borrowing from China "not only skills but also culture" in the sixth and seventh centuries, the Meiji modernizers emphasized by example the cultural habits of the West—clothing, food, buildings, law, education—as superior and progressive. The knowledge brought back from Europe and North America by the Japanese, or that of foreign experts working on contract in Japan, was systematically disseminated. Translations of Western books—both classical and contemporary—were not only made available in many areas but also lauded as valuable during these early Meiji decades.

At the same time, tradition was used creatively as a vehicle for stimulating desirable attitudes among the masses toward technical transformation. In order to promote a hard-working, loyal, and frugal workforce (one that would labor for low wages under difficult conditions), an Imperial Rescript on Education was promulgated in 1890, emphasizing self-discipline, traditional loyalties, constant knowledge seeking, and the idea of a common interest—all clothed in the garb of received tradition and imperial sanctions.

A Progressive Entrepreneurial Class

Overall, the use of the past had a crucial role to play in the type of modernization that Japan developed at the macro-level. The special role of the emperor, the Shinto myths, and the Samurai *bushido* ("way of the warrior") imbued the ideology of modernization in a uniquely Japanese direction. At the micro-level, the paternalism and business house traditions ensured a loyal and stable workforce. The merchants inherited from the Tokugawa period were a motley group; consisting of older merchants who had lent to the Daimyo and had lost 80 percent of their money, the new merchants of fortune, and the Meiji proteges *(Seisho)*. To create an independent, progressive, and responsible entrepreneurial group on the basis of such disparate members, the Meiji modernizers wrote, exhorted, publicized, and developed model projects in order to promote a new normative structure for businessmen. Honesty, independence, cooperative spirit, and social responsibility while making profits—all of these principles were adopted from the samurai code of ethics (*bushido*).

Organizational Restructuring

Technological, economic, social, political, and cultural changes were under way in the last third of the nineteenth century in North America and Europe, making possible and being made possible by a profound transformation of the institutional landscape. On the one hand, since the 1830s an explosive growth of numerous centralizing infrastructural technologies— railroads, postal system, telegraph—offered release from past constraints on coordination and control of various physically separate units of an enterprise and, indeed, allowed the integration of such large units, resulting in Alfred Chandler's "modern business enterprise." On the other hand, the emerging modern state began to expand its functional role beyond traditional concerns with law, order, and protection from foreign threats.

These links between the modern business enterprise, the modern state, and infrastructure technologies were effected by a number of generic organizational technologies. The organizational innovations provided the essential attributes of Max Weber's bureaucracy—the functional specialization, the critical techniques of coordination and control, the notion of salaried jobholder defined by roles and responsibilities. It was this stock of "on the shelf" social technology that the Meiji modernizers set out to borrow in order to support their industrialization program. Whereas most societies borrow organizational models from other societies, Meiji Japan was unusual in its readiness for institutional borrowing and the short time in which it completed such borrowing (Westney 1987).

As Eleanor Westney (1987) notes, Japan's borrowing focused on two types of organizations: the specific organization and the general organisation. Examples of the former include infrastructural organizations such as the telegraph (U.K.), the primary school and national banking systems (U.S.), and state organizations such as the army (Germany), judiciary (France), and navy (U.K.), which was fully transferred in its various components. However, in general organizations—such as factories, incorporated enterprises, and stock exchanges—the Japanese were more likely to adopt the primary organizational goal and some salient features required by the technology, and then to complete the organization with other structures existing within the Japanese environment and serving as functional equivalents in other aspects. In the latter case, there emerged several Japanese innovations during the process of institutional innovation (Westney, 1987; Lakshmanan, 1990).

Further, the Japanese developed a variety of indigenous institutions, such as the *Zaibatsu*, as the particular organization of the dualistic economy. Although the *Zaibatsu* has been deservedly criticized for a number of reasons, it not only sped up the transfer of new technologies and knowledge models but also provided for small and medium enterprises' access to technical, organizational, and marketing resources that they would not otherwise have had. Meanwhile, subcontracting networks (with secondary firms, subcontractors, and home-based part-time workers underneath), intercorporate stockholding, and long-term commitments and relationships provided a set of extramarket institutions that functioned to reduce risks and uncertainties, facilitated the flow of vital information, coordinated complex sequences of industrial and trading activities, and smoothed adjustments to fluctuations in the business cycle.

Material Infrastructures

Well known are the large-scale investments made in physical infrastructures during these early decades—railways, roads, coastal shipping, telegraph postal service, and so on—to facilitate rapid movement of materials, people and information necessary for industrialization.

In the area of human capital, Japan inherited from the Tokugawa period an educational and skill legacy, which in scale and quality was superior to that of most contemporary Western societies. This legacy encompassed not only impressive educational capabilities but also a range of skills in finance, commerce, and traditional production developed during proto-industrialization. It included a strong "will to learn" as well. The Meiji elites built on this by emphasizing universal basic education, expanding training opportunities in the newer production skills, and promoting higher education. Between 1860 and 1880, the number of schools doubled from 10,000 to 20,000; and the number of pupils, from 1 million to 2 million. The percentage of school-age children attending school increased from 40 percent in 1880 to 90 percent in 1900 (Beasley, 1990, pp. 94–95). By the end of the 19th century, the Japanese graduates had taken full advantage of the opportunities for higher education and peopled the upper reaches of the government bureaucracy and the large private enterprises.

Compensations and Bridges

How could the Meiji elites who sponsored economic, social, political, and cultural reforms of such vast scope encounter only relatively mild opposition? These reforms had abolished a variety of privileges for the feudal groups, separating the landlords from their historical functional role, their land, and their privileges. They also accomplished the civilianization of hundreds of thousands of samurai into urban jobholders; opened up the army, the navy, and the bureaucracy to all classes; and wiped out the older merchant class by declaring the freedom of commerce and the sweeping "adjustments" of federal debts (through which Tokugawa lenders lost 80 percent of their loans).

The reason lies in the fact that the Meiji sponsors of such social change provided not only a variety of economic and noneconomic compensations and cushions to these groups, whose historic economic and status assets were being devalued, but also a number of bridges between the old and new activities and institutions, thereby muting the resistance to change. Complicating the essential liquidation of the feudal class by the Meiji sponsors of social change was the fact that they were themselves members of that class, and that they saw the samurai class—educated, disciplined, achievement oriented, and nationalistic—as a potential vehicle of modernization. So the Meiji elites launched what in effect was a graduated structural adjustment program for the feudal group.

First, they viewed the continued support of a class with an obsolete function as a steadily declining and eventually terminating compensation tied to desired changes in economic and political behavior. As the Meiji elites replaced the decentralized feudal domains by a centralized systems of forty-eight prefectures in order to create a nation state, the daimyos were compensated in two ways: by their designation as imperial delegates in their former territories and by pensions that took them off the political scene.

Payments of the pensions consumed 38 percent of total government expenditures in 1871 and 29 percent in 1876. Indeed, the total costs of liquidating the feudal class were much higher (see Table 4.1). Consequently, all stipends were commuted in 1978 to interest-bearing government bonds (approximately 175 million yen), which were invested by most in business or in national banks after the passage of the 1876 Revised Banking Act.

The Meiji modernizers also lowered the costs of adjustment for the samurai class by reorienting their noneconomic incentives. One example is provided by the above-mentioned status designations of *daimyos* as imperial delegates in their former territories. Two other examples concern the appeal to the samurais' sense of public service and the linking of the special role of the emperor, the Shinto myths, and the *bushido* to the ideology of modernization. Such creative uses of the past not only lowered the adjustment costs but provided a normative motor for modernization of the samurai.

TABLE 4.1 Japanese Government Expenditures, 1868–1875

Fiscal Year	Liquidation of Old Feudal System*	All Other Expenditures
1868	82.4%	17.8%
1869	55.3%	44.7%
1870	45.2%	54.8%
1871	43.7%	56.3%
1872	45.8%	54.2%
1873	41.7%	58.3%
1874	52.1%	47.9%
1875	54.4%	45.6%

* The figures in this column denote compensations for feudal stipends, initial military operations against the Bakufu (mainly for 1868-1869), liquidations of feudal debts and subsidies, and loans to the feudal class for the start of businesses.
Source: Hirschmeier and Yui (1975), p. 81.

The decision by the Meiji modernizers to "unbundle" technology transfer from foreign capital in order to preserve economic and political independence also provided opportunities for creating bridges between the old and new activities in a way that reduced frictions and, indeed, created stakes for groups engaged in older activities in the emerging industrial society. Consider, for instance, the distinctive development of the Japanese "dualistic economy."

Given the late start, scarce capital, entrepreneurship, and the desire of the government to secure national defense and establish a communications system, an unbalanced growth pattern was inevitable. At one end of the industrial structure were large modern establishments, serving the modern (largely government and public-centered) internal market and aggressively expanding their foreign markets and large textile companies. At the other end was the mass of small and medium enterprises, left behind by the larger firms in terms of manpower, technology, and capital. In Japan, modernization did not wipe out these establishments over time as extensively as occured in many other countries; instead, it integrated them to a considerable degree in the growing modern sector and the consumer economy. These firms were exposed to modern technology either directly by the large firms or through the very special Japanese institution of trading companies (which scouted the world for appropriate technologies for these small firms). However, the majority, rather than being wiped out, were restructured by contact with Western technology and organization. The Japanese version of the dualistic economy, one may argue, preserved perhaps a greater portion of inherited traditional industrial capacity by introducing modern technology, organization, and power sources into small and medium enterprises, and linking them through various networks—thereby leading to the significant restructuring and evolutionary development of these enterprises and bridging them to the newer activities.

Thus it appears that the Meiji elites who sponsored industrialization were not guided by abstract principles of progress, by a drive to replace the old with the new. The social costs of such creative destruction could not be shrugged off; they had to be lowered, given the culture's emphasis on harmony. These sponsors of change had to be pragmatic and accommodate the old to the new. By encouraging the building of bridges between the traditional and the new, they creatively grafted old activities and institutions to the new.

Concluding Comments

The argument of this chapter can be summarized as follows. Technological progress both induces and is induced by pervasive social change. It not only represents a transformation in the material, structural, and cultural conditions of a society but also is made possible by such transformed conditions. The widespread diffusion of new technologies that characterize industrial societies, for instance, requires new modes of knowledge, new normative orientations, and new patterns of economic and institutional investment. As social actors emerge, proposing such new ethical models, economic philosophies, and innovations in social organization, they encounter friction or resistance from the groups whose modes of thought and action will be altered, transferred, or displaced. Since the groups in conflict have ideas and interests at stake as well as the means to fight for them, the outcomes are contingent. It is not clear, a priori, which of the contested models will prevail, what form they will take, and which groups will emerge to control these changes.

However, in the West, for at least two centuries prior to the emergence of the first industrial society, there has been an evolution toward an ambience favorable to technological progress. Through the presence of diverse groups, freedom to experiment, and the resources and incentives to generate technical and social innovations, the social environment in Western Europe has been tilting toward the activities of the sponsors of social change. The normative orientations toward enterprising and continuous innovations (in product, process, or organization) and a system for continuous change took root in Western Europe in such a way that industrial technologies arrived during the late eighteenth century in a relatively friendly social and economic environment. Still, the conflicts between the new and the old and between the "breakers" and "lamenters" were real, and the specific patterns and timing of the modernization varied according to the country's history, extramarket institutions, resource base, and the role of the government. In the case of late industrializers in other cultural realms, which had not undergone this normative and institutional preparation for industralization, the scope of necessary social change to facilitate the transition to industrial society and the corresponding level of resistance to such change were predictably large. The ability of the Meiji elites to establish in Japan new models of knowledge, new normative models, and new models of economic and institutional investment, all within the limited time they believed they had, was remarkable. It derived from (1) their clear-headed understanding of the discontinuity of the industrial transition and the need for material and nonmaterial infrastructures that would alter the basic context for thought and action; (2) their capacity for mobilizing and applying real and symbolic resources to the tasks at hand; (3) the reduction of frictions or resistance to change through a judicious mix of compensations, cushions, and bridges; and (4) their creativity in indigenizing the process of social change induced by technology through a blend of imported and Western meanings and social structures. Truly, the Meiji sponsors of Japanese industrialization were exceptional entrepreneurs of social change.

If one views the emergence of industrial society from a prior largely agricultural base as a shift in technical paradigm, there appear to be several paths or trajectories that can be and are taken within the new paradigm of modernity. Any such trajectory is a path of modernization taken by a society and is selective, path dependent, and cumulative. This is so because the way different societies marshal their material and nonmaterial resources for modernization is contingent on the outcomes of contests between groups that promote new modes of knowledge, ethics, and social structuring on the one hand and groups that oppose such models on the other. The Japanese transition to modernity outlined here provides an example of selective contrasts to the modernization experience in Europe and North America.

References

Beasley, W. G. *The Rise of Modern Japan*. London: St. Martin's Press, 1990.

Cardosa, F. H. "Dependency and Development in Latin America," *New Left Review,* 74; pp. 83-95; 1972.

Chenery, Hollis, and Ahluwalia, M. *Redistribution with Growth.* New York: Oxford University Press, 1974.

De Palma, Gabriel. "Dependency: A Formal Theory of Underdevelopment or a Methodology for the Analysis of Concrete Situations of Underdevelopment?" *World Development,* 6, pp. 881–924; 1978.

Dore, Ronald. *Flexible Rigidities: Industrial Policy and Structural Adjustment in the Japanese Economy 1970–1980.* Stanford, Calif.: Stanford University Press, 1986.

Dunn, Edgar S. *Economic and Social Development.* Baltimore: Johns Hopkins University Press, 1970.

Frank A. G. "Capitalism and Underdevelopment in Latin America: Historical Studies of Chile and Brasil," *New York Monthly Review Press,* 1967.

Gerschenkron, Alexander. *Economic Backwardness in Historical Perspective.* Cambridge; Mass.: Harvard University Press, 1966.

Giddens, Anthony. *Central Problems in Social Theory.* London: Macmillan, 1979.

Hirschmeier, J., and Yui, T. *The Development of Japanese Business 1600–1973.* Cambridge; Mass.: Harvard University Press, 1975.

Lakshmanan, T. R. "Infrastructure and Economic Transformation."; In *Advances in Spatial Theory and Dynamics*; eds. Åke E. Andersson, P. Nijkamp, D. Batten, and B. Johansson. New York: North Holland Press, 1989.

———. "The Social Context of Technological Progress in Japan." Paper presented at the European Regional Science Congress, Istanbul; August 27–31 1990.

Morishima, Michio. *Why Has Japan Succeeded? Western technology and the Japanese Ethics.* New York: Cambridge University Press, 1982.

Rosenberg, Nathan, and Birdzill, Nancy. *How the West Got Rich.* New York: Basic Books, 1986.

Sen, Amartya. "The Concept of Development," in H. Chenery and T. N. Srinivasan, Eds., *Handbook of Development Economics,* Vol. 1. Amsterdam: Elsevier Science Publishers, 1988.

Smith, Robert J. *Japanese Society: Tradition, Self, and Social Order.* Cambridge: Cambridge University Press, 1986.

Streeten, Paul, et al. *First Things First: Meeting Basic Human Needs in Developing Countries.* New York: Oxford University Press, 1981.

Tawney, R. H. *The Sickness of an Acquisitive Society.* London: Fabian Society / G. Allen and Urwin, 1920.

Touraine, Alain. *The Post-Industrial Society. Tomorrow's Social History: Classes, Conflicts and Culture in the Programmed Society.* New York: Random House, 1971.

———. *Return of The Actor: Social Theory in Post-Industrial Society.* Minneapolis: University of Minnesota Press, 1988.

Vickers, Geoffrey. *Making Institutions Work.* London: Associated Business Programmes, 1973.

Viner, J. "Religous Thought and Economic Society." Unfinished work. Durham,

Viner, J. "Religous Thought and Economic Society." Unfinished work. Durham, N.C.: Duke University Press, 1978.

Weber, Max. *The Protestant Ethic and the Spirit of Capitalism.* Translated by Talcot Parsons. New York: Scribner, 1930.

Westney, Eleanor D. *Imitation and Innovation: The Transfer of Western Organizational Patterns to Meiji Japan.* Cambridge, Mass: Harvard University Press, 1987.

5

Inertia and Development Models

Georg Sørensen

This chapter is about inertia in relation to models of development.[1] Terminology first. I use the *Encyclopedia Britannica's* (1985:306) definition of *inertia:* "Property of body by virtue of which it opposes any agency that attempts to put it in motion or, if it is moving, to change the magnitude or direction of its velocity." A *model of development* contains three components: (a) a concept of development defining the goal of the development process (i.e., clarifying the notion of development itself); (b) a strategy of development, concerning the guidelines for action intended to move the process of development along; and (c), a theory of development, meaning a set of hypotheses regarding conditions, relationships, and structures that are held to be important for the process of development. Furthermore, the *processes of development* discussed here concern the level of societies, though I also refer to specific groups and individuals. The distinctions among the three dimensions of development models are purely analytical, inasmuch as they are closely interrelated. The formulation of a strategy requires theoretical considerations that must also include a notion of the goal—namely, the concept of development.

I shall concentrate on inertia in relation to the concept of development, on the one hand, and concrete examples of development experiences, on the other.

Inertia and the Concept of Development

Let me begin with the concept of development. The idea of development is one of the oldest and most powerful Western ideas (Hettne, 1990, p. 39). It is rooted in antiquity, but its intimate connection with the idea of growth came later. The notion of growth gained prominence in the thinking on sociology of evolution connected with the strong development optimism of the nineteenth century. This thinking, in turn, can be understood only against the background of industrial capitalism's breakthrough in Western Europe.

Herbert Spencer (1820–1903) is a typical representative of the evolutionary optimism that was the foundation of this thinking. His ideas on society's development are based on an analogy between the biological organism and human society. Modern, industrial society, he claimed, is the product of a process of evolution just as the human organism is the superior product of a process of organic evolution. As G. Duncan Mitchell explains it,

> the history of both organic and superorganic (social) life is a process of development and this development involves an increase in both quantity and complexity. Thus, just as the earliest forms of organic life were unicellular and simple in structure, so early mankind lived in a few simple isolated groups or hordes; and just as later forms of organic life display differentiation and multiplicity in both structure and function, so do more recent forms of social life. (1968, p. 24).

Deeply ingrained in Western thinking, then, is the notion that the central core of development is growth. The growth, moreover, is in both quantity and quality, and it takes a specific direction: from the simple toward the complex (i.e., toward increasing differentiation and sophistication).

The development problem appeared on the agenda in the modern sense after World War II, when a large number of countries had gained independence and voiced their demands for development in international organizations that were founded, especially in the United Nations' system. The dominant trend of development thinking established in this period is called modernization theory. It built directly on the considerations outlined earlier. Development basically means growth, leading toward an increasingly complex society. It is logical, then, to describe modern, complex, industrial societies as developed and the more agrarian, traditional, backward societies in Africa, Asia, and Latin American as less developed or nondeveloped. Against this background, it was not only for practical reasons that the core indicator of development came to be the level and ratio of economic growth. In a very real sense, economic growth expresses the modern Western meaning of what development is. The World Bank continues to produce its annual ranking of countries on the basis of gross domestic product (GDP) per capita (World Bank, 1991). The Danish law concerning development aid, for example, defines the achievement of economic growth as the basic means of supporting the progress of recipient countries. Moreover, this is not something the leaders of developing countries are forced to accept; rather, it is a true reflection of their own demands for growth, given that rapid industrialization leads toward modern, Western society (cf. Barraclough, 1980).

My contention is that the idea of growth, with a special emphasis on economic growth in a process of modernization, has, to a frightening extent, captured our idea of what development is. There are alternative ideas about development, and the notion itself is subject to change, as I shall demonstrate in a moment. But the Western concept of development (the dominant concept, on a world scale) is so deeply imbued with the idea of growth toward modernization that it has been almost totally able to "oppose any agency" that has attempted to "change the magnitude or direction" of this understanding. In a very real sense, then, the notion of development is subject to powerful forces of inertia. It follows that inertia can be present at different degrees of intensity. The inertia connected with the understanding of development as growth toward modernization must be one of the strongest examples of almost irresistible inertia, because it is welded into the core structure of our societies themselves. Growth is progress and development, for us and for everybody else.

The power of the idea is manifested in the thinking of writers normally viewed as critical of modern capitalist society. Karl Marx, for example, considered the process of growth away from traditional and toward modern capitalist society to be a progressive process of real development. The road toward modernization in a process of growth was the road to take for backward countries; "the industrially more developed country showed to the less developed the image of its own future." So, although the English plundered and mistreated India, they also crushed the structures of traditional society and paved the way for a modern one. Accordingly, their efforts, even from an Indian viewpoint, must be considered progressive; with traditional community cleared away, the stage was set for the growth of modern society. In a similar vein, Lenin regarded capitalist growth in Russia as a progressive process of development.

It is true that the critical theorists of underdevelopment who came forward in the 1960s and 1970s turned strongly against the idea that development could be achieved through contacts with the already modernized Western countries. In fact, they declared that the opposite is true—that the modern Western countries have contributed to the underdevelopment of the countries of Asia, Africa, and Latin America. Therefore, the poor countries must break off all connections with the Western exploiters and seek their own process of development. The point here, however, is that the goal set for this autonomous process was exactly the same as the one laid down in the Western notion of development: a process of (economic) growth, leading toward a modern industrial society. The inertia of the Western concept of development prevailed again, as it did in the development goals set by the planned economies of the former Soviet Union, Eastern Europe, China, and North Korea. In all these cases, the aim was never to conceive of a development alternative to the Western one; rather, it was to catch up and, ideally, to beat the Western competitors at their own game: industrial growth and modernization. Therefore, the present changes in Eastern Europe and the former Soviet Union do not represent a retreat from the old goals of growth and modernization but, instead, reflect a concession that these goals could not be reached by totalitarian governments running centrally planned economies. The whole idea of attempting a political and economic transformation of Eastern Europe is to unleash the forces of growth toward modernization in a more effective manner than under the old system.

Some modifications are in order, of course. There have always been objections and protests against growth and modernization. These have come most forcefully from groups threatened with extinction by the growth process: small farmers, small-scale business people, artisans, and so on, often supported by various intellectuals. In an excellent book written some years ago, Gavin Kitching (1982) identified the thinking of these diverse groups as reflecting conservative romanticism as well as utopian socialism, anarchism and populism. The strongest current expression of criticism toward the growth-modernization ideal as the given goal of development comes from the Green movement and other ecology movements concerned with the negative effects of growth on the environment and our limited resources. This criticism has been combined with an increasing awareness of the fact that not all growth processes are equally capable of providing welfare gains for people; indeed, growth can be more or less oriented toward human development (UNDP, 1991).

Yet these countermovements are only slowly carving out a place in a model of development still massively dominated by the growth-modernization perspective. It is symptomatic in this context that the much talked about Brundtland Report contains merely a description of environmental problems, while avoiding a serious analysis of causes and implications (cf. Trainer, 1990). From a global perspective, environmental and resource problems are still very much the topic of conferences and academic gatherings, even as the vast majority of governments pursue the old game of growth-modernization. These governments included that of Ms. Brundtland herself.

Let me try to summarize the aforementioned considerations from the perspective of inertia. Several dilemmas present themselves against the background of a dominant goal of development committed to growth-modernization. The most important one, of course, is that the forces of inertia are so strong that the dominant conception of development will stay in place or change only very slowly, such that the environmental and resource damages to the planet and its ecosystems become irreparable. Several commentators clearly believe that this is already the case.

At the same time, we have seen environmental concerns climb high on the agenda of certain rich countries. This circumstance is a step forward; but when the pendulum starts swinging toward environmental concerns, its velocity, through forces of inertia, can also cause it swing all the way to the other extreme. For example, it was the vivid image of a basket of dead lobsters during a Danish television news program that rapidly paved the way for committing 10 billion kroner ($1.8 billion) for improving the quality of Danish waters, a move that was later questioned. Yet I admit that there are not many examples of this type. The dominant thrust is the continuation of the growth-modernization model of development.

From a more general perspective, it seems that forces of inertia prevent us from achieving a flexible, dialectical notion of proper development goals that can provide a framework for more concrete measures towards development. Either we stay at one extreme, holding onto the dominant, deep-seated notion of development as growth-modernization, or, when the pendulum starts to swing, we move so fast toward the other extreme that the productive interplay between different types of conflicting goals becomes impossible. This may be one of the gravest problems in development, springing from the forces of inertia.

Inertia and Development Experiences

Development experiences are concrete examples of development. As such they combine the three elements of a development model: (1) a goal of development is pursued through (2) a strategy, in the framework of (3) a theory of development. Whereas the goal of development has remained relatively fixed in the growth-modernization formula, the theories and strategies have been subject to many changes over time. This is not surprising. There are so many elements important to processes of development (economic, political, social, cultural, etc.) on so many different levels (global, regional, national, local) that they are bound to be combined or interpreted in different ways by scholars and practitioners.

Space does not allow an exposé of dominant theories and strategies. I wish to emphasize merely one point in relation to inertia. It has to do with the fact that, at any given time, there is a dominant "development dogma"— that is, a set of recommendations that dominate theorizing and strategic thinking on development. Such dominance is never total, of course. But important international organizations, the media, and leading politicians can come together in setting the agenda for discussion of theories and strategies of development. The dominant dogma in the 1980s was produced primarily by neoclassical economists, promoted by Thatcher and Reagan, and put to use in developing countries by the International Monetary Fund (IMF) and the World Bank. This neoliberal thinking stressed the advantages of market forces as opposed to state intervention in the economy. It also opted for integration of developing countries in the world market and gave high priority to industrial growth in the private sector. (For a full treatment of this topic, see Toye, 1987.)

I have argued elsewhere that this dominant dogma is in the process of being succeeded by a different one in which the earlier neoliberal emphasis on market and private entrepreneurship is being revised, toward a "Social Democratic" vision of combining state and market, growth and welfare (Sørensen, 1991). What must be stressed here, however, is the mere existence of such dominant development dogmas. The point is that at any given time, some theories and strategies of development are more prone to be followed than others. These are the windows of opportunity that are open, while other windows, based on alternative considerations, are closed. Especially in relation to smaller, very poor countries in the developing world, this insight is an important one, given that poor countries are highly dependent on support from the international system and, hence, more directly subjected to the dominant ways of thinking. Due to forces of inertia, such dominant ways of thinking change only slowly and gradually. In the meantime, the people in these countries are exposed to their effects. In Africa, for example, market-oriented measures have been undertaken primarily through IMF and World Bank adjustment and restructuring programs. In 1988, the World Bank chief economist for Africa stated, "We did not think that the human costs of these programs could be so great and economic gains so slow in coming" (cf. Bienefeld, 1989).

Yet it would be wrong to argue that concrete development processes are shaped only or even primarily by systematic theoretical and strategic con-siderations. The decisive factor behind a concrete path of development is the constellation of social forces upon which it is based (cf. Sørensen, 1986). For example, in the so-called classic division of labor, the developing coun-tries specialized in exports of raw materials and agrarian products while

Georg Sørensen

importing manufactures from the industrialized countries. But their political leaders did not follow this path primarily because they had read the theoretical justification for it in the theory of comparative economic advantage formulated by David Ricardo. Rather, they followed it because the societies and states of these countries were dominated by exporters of raw materials and agrarian produce that stood to gain from pursuing the classic division of labor. When they later changed their strategies in the direction of promoting industrialization under their own control, it reflected a changing constellation of social forces, with a rising local bourgeoisie increasing its political influence. Therefore, when we look at concrete development experiences, it is relevant to ask: What is the constellation of social forces behind the strategy? Who has major political influence? The answers to these questions will help us understand how development efforts are designed and implemented. In this sense, then, the forces of inertia are located less in the dominant development dogma than in prevailing constellations of social forces and their resulting political influence. It is a reasonable hypothesis that radical changes of development paths involve radical shifts in power relations and the forming of new coalitions in power. In many developing countries, old dominant elites have been in power for decades—another major reason for resisting inertia when alternative, more effective models of development are being formulated.

The crucial agency for taking measures towards development is the state, a point that Gunnar Myrdal (1968) and many others have stressed in their analyses. In other words, state apparatuses and the governments that command them are the most important agents for the promotion of development in any country. The point in relation to inertia is that the elite who controls the state can enjoy more or less autonomy depending on the dominant social forces of society mentioned earlier. There is an inverse relationship between autonomy and inertia: A low degree of autonomy means a high degree of inertia. State leaders can make changes only after prolonged negotiations and compromise with different groups in society, and the process itself may tend to water down these changes such that they become piecemeal adjustments rather than swift and radical shifts in direction. And vice versa: A high degree of autonomy means a low degree of inertia. State leaders can effectuate radical and far-reaching changes rapidly, without lengthy prior consultation and compromise.

Scholars and others who have sometimes assumed that democratic governments are much less autonomous than authoritarian ones (cf. Sørensen, 1991) tend to view democratic governments as less effective in terms of promoting economic development: Subject as these governments are to strong forces of inertia, they are unable to bring about the swift changes that are often needed in order to drive a process of development forward. Specifically, democracies are charged with being too dependent on short-term considerations having to do with winning the next election. Therefore, so runs this reasoning, they are not able to properly consider the long-term development priorities of their country.

Such logic, if it claims general validity, is much too crude. First, there are so many different types of both authoritarian and democratic systems that is impossible to make the general claim that autonomy pertains to all the cases in one category but to none in the other. Second, there is no straightforward relationship between presence or lack of autonomy (or inertia) and the results of development. In order to discuss the complexities of this relationship, I will focus on three cases—one involving a low degree of autonomy and a correspondingly high degree of inertia (India), another involving a very high degree of autonomy and a low degree of inertia (China), and a third involving a rather distinctive combination of autonomy and inertia (Taiwan).

India: The Forces of Inertia

India is by no means a development failure, according to the conventional indicators. Its rates of growth in industry (around 5 percent per annum) and structural change are respectable. It is capable of producing almost the entire palette of industrial goods itself, including capital goods. Although there has been less growth in agriculture, the so-called Green Revolution, which has yielded improved varieties of seeds and more intensive use of irrigation, fertilizer, and pesticides, has made India almost self-sufficient in terms of food.

Yet these achievements have not meant substantial improvements for the poverty-stricken masses in India. Around 1950, when the population was 360 million, about half were subsisting below the poverty line. Forty years later, in 1990, the situation was unchanged: 51 percent of the rural population and 40 percent of the urban population were still below the poverty line. Given a population increase to some 820 million, this amounts to 394 million people in poverty. According to the indicators of human development employed by the United Nations (life expectancy at birth, adult literacy, and purchase power), India maintains a low rank in human development (UNDP, 1990).

Why has there not been more improvement for the poor majority? We may find a possible explanation by looking at the structure of socioeconomic and political power that forms the basis of the democratic governments in India. Three groups hold dominant influence in politics: the urban professionals, the bourgeoisie in industry and trade, and the rural landowning elite. It is these groups, constituting about 20 percent of the population, that have benefited from government policies by defining the boundaries within which they could take place. Together, they form a dominant coalition, capable of restricting the autonomy of governments in the sense that development policies have never veered beyond what the coalition finds acceptable. Meanwhile, the masses are simply too unorganized, divided, and politically weak to radically change this state of affairs (Sørensen, 1991).

Inertia in India, then, refers to the incapacity as well as the unwillingness of governments to radically break the narrow limits of maneuver defined by the dominant coalition and thereby implement more radical reforms that could improve the welfare of the poor by attacking elite privileges. However, if there is a constructive side of inertia in India, it has to do with another aspect of the constraining effects of the dominant coalition. India's policies have not tended to go astray, leading to unforeseen, catastrophic situations. For example, there have been no severe famines in India since independence. The same cannot be said for China, the country to which we now turn.

China: The Absence of Inertia

Whereas in India power was transferred from the English in a slow process of bargaining to a dominant coalition who subscribed at least in principle to the same liberal democratic ideals as the colonial masters, the situation in China was quite different. Four decades of political instability and civil war preceded the communist victory in 1949. In effect, Mao's triumph left the communists a very small group of leaders headed by Mao, in complete control of the huge country. During the civil war, the party's responsiveness to the masses had been a condition of its survival in an otherwise hostile environment. But victory meant an entirely different relationship. As B. Womack described it:

> There is no legitimate opposition; there is hardly any legitimate interest articulation. . . . The masses must rely on the party and its cadres, but the party and its cadres are no longer in a position of direct, reciprocal dependence and vulnerability. The remaining clout of the masses is reduced to their

discretion as producers, consumers, and risktaking resisters. The masses have become the fish, and the party controls the water. (1987, p. 498n).

Given the extremely hierarchical nature of the communist party organization, whereby supreme power is concentrated at the top, we get a situation where this leadership enjoys an extraordinarily high degree of autonomy. In other words, the leaders are free from forces of inertia in the sense that they can pursue almost any kind of policy, unconstrained by forces in society. The freedom from inertia is not complete, of course; the leadership must eventually deliver a measure of economic development in the form of welfare gains or it risks paving the way for rival coalitions. Yet in both the short and medium runs, the constraints on the leadership are almost nonexistent.

This situation hardly worried the peasants and workers in 1949. The communists had gained massive popular support and prestige through the reforms undertaken in the liberated areas. And it was a high priority for the communists to improve the life of the poor through rapid economic development.

This is what happened. The leadership used its room of manoeuvre to pursue radical economic reforms and redistribution of wealth. Land reform alone made a dramatic difference for the rural poor. According to one estimate, the share of total income going to the poorest fifth of the rural households almost doubled, from 6 to 11 percent (Riskin, 1987, p. 235). Overall, China has experienced high rates of economic growth, especially in industry. After the initial reforms, agriculture was left behind, severely pressured to deliver resources for industrialization; it was only after the Deng reforms in the late 1970s that it came back on steam.

The welfare achievements in China have been remarkable. Starting from a level quite similar to that of India, China has made much more progress, as can be seen in Table 5.1. Poverty has not been eradicated, but China has a much smaller number of people in poverty than does India, in both absolute and relative terms. That number, according to the best available estimates, is 12 to 15 percent of the population (Sørensen, 1991, ch. 3).

Yet there are two more sides to the picture. First, growth had a higher priority than consumption in Mao's time. In the early 1950s, resources for investment and growth could be procured without hurting mass consumption, because the surplus that earlier had accrued to landlords and bourgeoisie was now made available for investment. But during the twenty years between 1956 and 1976, high rates of growth were possible only through curbing mass consumption to an average level—a level not very much above subsistence in the rural areas (Ka and Selden, 1986,

Georg Sørensen

p. 1300n). That this outcome could be combined with the achievements shown in Table 5.1 is due to the fact that certain direct measures of health care and education improved the overall welfare situation. But by no means did the average peasant become affluent.

Second, and worse, the lack of forces of inertia allowed the communist leadership not only to pursue good policies of economic and social progress but also to commit horrifying mistakes with devastating consequences for the population. This is what happened during the so-called Great Leap Forward in 1958-1961, which was supposed to propel China into the ranks of highly industrialized nations in a few years' time. Instead, chaos was the result: Grain production dropped, first by fifteen percent, then by 25 percent, and ultimately by 50 percent. The consequence was a disastrous famine that killed several million people, possibly as many as 30 million (Ashton, 1984, p. 634).

It is not that Mao's intentions were bad; indeed, the Leap strategy was intended to correct some of the problems in the earlier development plans. Rather, the obstacles were severely underestimated and in what can only be seen as sheer arrogance of power, Mao held onto the strategy even after it was proven wrong, thereby adding to the ensuing disaster. Why this stubborn attitude? The objective was to avoid showing weakness in the face of Marshal Peng Dehuai's criticism of the Leap policies (Riskin, 1987, p. 276).

There were good intentions behind the Cultural Revolution as well: primarily the attack on elitism and the rule of experts. But, again, chaos and anarchy reigned and nearly everyone was suspected of "following the

TABLE 5.1. Welfare in China and India

Indicators	China	India
Life expectancy at birth (in years), 1990	70.1	59.1
Adult literacy rate (in percentages), 1985	68.2	44.1
Mean years of schooling, 1980	4.8	2.2
Human Development Index (HDI) Rank[a]	0.614	0.308

a) The HDI takes three indicators into account: life expectancy, education, and income. For each indicator, a worldwide maximum (1) and a minimum (0) are identified, and a given country is then ranked according to its position. The combined average of the three positions is the HDI; the closer it is to 1, the better the ranking.

Source: Figures from UNDP, (1991), pp. 88–91, 120.

capitalist road." The harassment and utter humiliation of as many as 100 million people were the results (Thurston, 1984–1985).

In summary, absence of inertia can pave the way for radical policies, with significant improvements for the population. It can also lead to mistaken policies, putting the population at the complete mercy of its leaders. If we look at the situation just in terms of health and poverty, China is easily the winner in comparison with India: "Every eight years or so more people die in India because of its higher regular death rate than died in China in the gigantic famine of 1958–61. India seems to manage to fill its cupboard with more skeletons every eight years than China put there in its years of shame" (Drèze and Sen, 1989, p. 215). But from a wider perspective, in terms of the value placed on the much higher respect for basic civil and political rights in India, the final judgment is perhaps less easy to ascertain. In any case, India demonstrates the negative effects of a high degree of inertia constraining the leadership, whereas China demonstrates both the negative and positive effects of the opposite situation.

Taiwan: Autonomy and Inertia Combined

The India/China comparison gives us an interesting reason to look for possible cases in which the positive effects of high autonomy and low inertia are combined with mechanisms that inhibit the negative effects, in the form of policy excesses and destructive mistakes. Is it possible to identify such cases? Taiwan seems to be a good candidate. The Guomindang regime, which took over from the Japanese colonialists after World War II, came from the mainland, defeated by Mao's communists. Because its leaders had no ties of obligation or allegiance toward vested interests on the island, they were free to pursue the goal of rapid economic development.

This they accomplished through a model that avoided the major problems that had created setbacks for the Maoist strategy (Gold, 1986). Agrarian reforms laid the basis for a system of highly productive family farming which provided an economic surplus for industrialization. Now there was no Stalinist overemphasis on heavy industry that continued to plague the industrial structure on the mainland. On the contrary: Emphasis was on light industry, which was labor intensive. Rapid growth absorbed the expanding labor force. The first phase of inward-looking, import-substituting industrialization gave way to a second phase of export-oriented industrial growth. Interaction with the world market, combined with support from the state, helped industries to gradually upgrade and increase competitiveness. Indeed, there was a productive interplay between an active state which guided and controlled the economy, and private-led economic activities. State and market forces were combined, in contrast to the Chinese attempt to strangle all market mechanisms.

The results have been remarkable, not only in terms of high and stable rates of economic growth but also in terms of welfare achievements. Moreover, the achievements are relatively equally distributed; the model has successfully mix fast growth with a high degree of equality. As a result, there is only a small fraction of the population below the poverty line. According to one recent estimate, this fraction amounts to less than 3 percent (Yun-peng Chu, 1987).

Why did the lack of inertia constraining the Guomindang regime under Jiang Kaishek lead to one success in development after another? Why was there no occurrence of the mistakes and setbacks seen on the mainland? These questions are even more puzzling when one considers the senseless rule of Guomindang on the mainland. The communists prevailed in the civil war, not primarily because of their own strength but, rather, because of the incompetent, corrupt, and oppressive rule of the Guomindang.

Several factors helped set Jiang Kaishek and his party on an efficient course toward development. It was clear that in order to win legitimacy in Taiwan, the islanders had to be given an incentive to support Guomindang. A fundamental cleaning of the party was undertaken, which cut away some of the worst rascals. The humiliating defeat to the communists was an incentive to the leaders themselves to organize a more efficient operation.

Even so, it is difficult to see why the Guomindang, left to its own resources, did not commit severe mistakes and indulge in failing policies. This did not happen because, in the case of Guomindang in Taiwan, the high autonomy/low inertia situation in relation to domestic social forces was combined with an extremely high external dependence on an entity that could constrain and moderate the movement and keep it firmly on the right track.

This entity was the United States, of course. Keenly interested in containing the spread of communism, the United States was preoccupied with promoting a successful showcase of noncommunist development right at China's doorstep. That meant ample economic aid for Taiwan, but noneconomic aid was even more important. The United States provided the experts who could help promote economic development. It promoted agrarian reform in Taiwan at an early stage, and it was instrumental in keeping a place for market forces and private capital in the development model (Jiang would have preferred complete state control) as well as for the turn toward export orientation in the second phase of development.

On all important issues connected with economic development, the United States provided advice that was very difficult for the regime to disregard.

This is not to say that the United States was solely responsible for Taiwan's success in economic development. But it was the decisive factor in determining why the absence of constraining forces of inertia on the Jiang Kaishek regime did not lead to the major mistakes and setbacks that we saw in China's case. It seems clear that Taiwan demanded a rather distinctive mix of internal and external conditions in order to achieve the fortunate combination of autonomy and inertia that was instrumental in propelling the country's development. Only Japan and South Korea have experienced similar conditions. The other countries in Southeast Asia, not to mention the countries in Africa and Latin America, have few chances of replicating this pattern.

In this connection, the prospects for other countries are less bright. India, for instance, will be left to the highly constraining forces of inertia or, at the other extreme, to completely unconstrained rulers whose policies may be disastrous. As the Uganda of Idi Amin, the Central African Republic of Jean Bedel Bokassa, and the Campuchea of Pol Pot demonstrate, policies gone astray can have consequences even more serious than China experienced.

The Inertia of Strategies of Development

The foregoing analysis has focused on the presence or absence of inertia in one significant respect: the degree of autonomy of state leaders to pursue development objectives. The possibilities but also the dangers inherent in a low inertia/high autonomy situation have been highlighted. Yet it is necessary to mention an additional element that complicates the picture. It has to do with the inertia built into specific strategies of development. State leaders may be unconstrained in relation to social forces of society and, hence, free from inertia, as we saw in the case of China. Yet forces of inertia may also be present in terms of the strategy of development that the leadership sets out to follow.

For example, Mao and his followers were confirmed communists. So it was hardly surprising that they set out to imitate the Soviet experience, formulating a firmly Stalinist plan with emphasis on heavy industry and growth in the "modern" sector. Furthermore, China had been in a situation of international isolation since the war in Korea. Aid could come only from the Soviet Union. In addition, the USSR was at the time regarded as a successful model of socialist development, as "a reliable guide to drawing a backward, agrarian nation into the ranks of industrial strength" (Rosenberg and Young, 1982, p. 233).

But this way of copying the Soviet model had to involve reproducing the same problems that plagued the Soviet Union: (1) an overemphasis on heavy industrial accumulation, such that consumer goods are few in number and of poor quality; (2) a centrally planned system that is directly inimical to innovation and flexibility and offers no incentives for ordinary workers to increase their efforts; and (3) a complete lack of concern for environmental issues. Mao was incapable of improving this system. At the same time, he would not allow any role for market forces whatsoever, because he viewed such forces as capitalism sneaking in through the back door. In this way, the highly autonomous Chinese leadership was bound by other forces of inertia, set in the model of development that they attempted to emulate. These forces of inertia continue to work at the political level, even though gradual changes have taken place since Mao's death. The present leaders remain basically committed to the old model of centralized communist rule, not least because dramatic changes would threaten their own positions of power.

These forces of inertia, tied in with the strategy of development, are also present in the cases of India and Taiwan, but in a less pronounced way: Those countries never committed themselves to one specific strategy as dramatically as did China. India under Nehru actually saw China as a desirable model, but in liberal democratic India he was not free to pursue it in a way that might have threatened the dominant coalition. The irony, then, is that the high degree of inertia that narrowed the freedom of maneuver of Nehru and other Indian leaders, due to their "obligations" to the dominant coalition, freed them to some extent from the constraining strategic inertia that would have resulted from a narrow-minded emulation of the Chinese model of development.

Strategic inertia in India and Taiwan has thus been less constraining than in China. On the other hand, both countries have pursued versions of capitalist growth models in a context of a large measure of state involvement. Both are experiencing problems at the moment. India has environmental problems related to the Green Revolution strategy in agriculture, with its emphasis on irrigation and pesticides, and efficiency problems related to the large sector of state enterprises. Taiwan has severely increasing environmental problems of its own, after several decades of rapid industrial growth without environmental concerns. Furthermore, Taiwan is also facing increasing competition in its continued role as a producer of low labor-cost consumer goods for the world market (Bello and Rosenfeld, 1990).

Conclusion

Inertia affects development issues in several ways. It is instrumental in the continued maintenance of an understanding of development itself as growth toward modernization. If and when this understanding is basically questioned, inertia can play another trick, causing the pendulum to swing forcefully toward the other extreme. In short, forces of inertia impede the attainment of a dialectical, flexible notion of various development goals that would be a more proper framework for concrete efforts.

Moreover, inertia plays a role in the competition between different theories and strategies of development. At any given time, there are certain dominant development dogmas that are prone to be followed, while others are not. The dominant ways of thinking may not reflect the best solutions for the developing countries, but forces of inertia mean that they change only gradually.

In many developing countries the old, dominant elites constitute a force inertia that stands in the way of implementing other, more effective models of development. This factor is tied in with another one: the degree of autonomy of those in charge of the state. Three cases were outlined earlier: India, with low autonomy/high inertia; China, with high autonomy/low inertia; and Taiwan, with a fortunate combination of autonomy and inertia. As I suggested, the Taiwanese case was rather distinctive and not likely to be replicated beyond a few other East Asian countries. Accordingly, the vast majority of countries are left with the unpleasant alternative of very high inertia, blocking radical reforms capable of making improvements for the poor majority, or very low inertia, with the risk of policies leading to disasters.

Against this background it may be constructive for states that wish to support the development efforts of poor countries to think of ways of achieving better combinations of autonomy and inertia, for example, by supporting projects in India that lead to a strengthening of popular influence and a corresponding weakening of the position of elite forces. But this is clearly a long-term process. Ironically, the aid from Scandinavia is subject to its own forces of inertia because of the vested interests of business communities. Furthermore, I noted that even in the case of political leaders with high autonomy, other forces of strategic inertia may be at work. It is impossible to escape inertia in any aspect of development, the most one can hope for is to better learn how to play on the forces of inertia in order to secure the best conditions for development progress.

114   *Georg Sørensen*

References

Ashton, B., et al. "Famine in China," *Population and Development Review*, vol. 10, no. 4, December 1984.
Barraclough, Geoffrey. "Worlds Apart: Untimely Thoughts on Development and Development Strategies." *IDS Discussion Paper 152*, Sussex: Institute of Development Studies. 1980.
Bello, Walden, and Rosenfeld, Stephanie. "Dragons in Distress: The Crisis of the NICs," *World Policy Journal*, vol. 7, no. 3, pp. 431–469. 1990.
Bienefeld, Manfred. "The Lessons of History and the Developing World," *Monthly Review*, vol. 41, no. 3, July-August, pp. 9–42, 1989.
Drèze, Jean & Sen, A. *Hunger and Public Action*. Oxford: Clarendon Press, 1989.
Encyclopædia Britannica vol. 6, Chicago: The University of Chicago, 1985.
Gold, Thomas B. *State and Society in the Taiwan Miracle*. New York, 1986
Hettne, Björn. *Development Theory and the Three Worlds*. Burnt Mill, Harlow: Longman Scientific and Technical, 1990.
Ka, C. & Selden, M. "Original Accumulation, Equity and Late Industrialization: The Cases of Socialist China and Capitalist Taiwan". *World Development*, vol. 14, no. 10–11, 1986.
Kitching, Gavin. *Development and Underdevelopment in Historical Perspective: Populism, Nationalism and Industrialization*. London and New York: Methuen, 1982.
Mitchell, G. Duncan. *A Hundred Years of Sociology*. London, 1968.
Myrdal, Gunnar. *Asian Drama: An Inquiry into the Poverty of Nations*, vols. 1-3. New York: Pantheon, 1986.
Riskin, Carl. *China's Political Economy: The Quest for Development Since 1949*. Oxford: Oxford University Press, 1987.
Rosenberg, W. G., and Young, M. B. *Transforming Russia and China: Revolutionary Struggle in the Twentieth Century*. New York, 1982.
Sørensen, Georg. *Udviklingsteori og den tredje verden* (Development Theory and the Third World). Aalborg: Aalborg University Press, 1986.
———. *Democracy, Dictatorship and Development: Economic Development in Selected Regimes of the Third World*. London/New York: Macmillan and St. Martin's Press, 1991.
Thurston, Anne F. "Victims of China's Cultural Revolution: The Invisible Wounds," *Pacific Affairs*, vol. 57, no. 4, 1984-1985.
Toye, John. *Dilemmas of Development: Reflections on the Counterrevolution in Development Theory and Policy*. Oxford: Basil Blackwell, 1987.
Trainer, Ted. "A Rejection of the Brundtland Report," *ifda dossier*, no. 77, May-June, pp. 71-85, 1990.
UNDP (United Nations Development Project). *Human Development Report 1991*. New York/Oxford: Oxford University Press, 1991.
———. *Human Development Report 1990*. New York/Oxford: Oxford University Press, 1990.
Womack, B. "The Party and the People: Revolutionary and Postrevolutionary Politics in China and Vietnam," *World Politics*, vol. 29, no. 4, July, 1987.

World Bank. *World Development Report 1991*. Washington: World Bank, 1991.

Yun-peng Chu. "Taiwan's Poverty: Decomposition and Policy Simulations." Paper presented at the Conference on Economic Development and Social Welfare in Taiwan, Taipei, 1987.

PART FOUR

Rationality in the Marketplace

6

Friction in Economics

Keith Griffin

The core of late-twentieth-century economics consists of a set of assumptions that, taken together and seen as a whole, constitute an image or vision of a self-regulating system in which friction and inertia are largely absent. Friction in both senses of the word is generally ignored: there is neither dissension nor resistance to relative motion. The economic system adjusts smoothly to disturbances; markets clear instantaneously; competition ensures that resources are used efficiently. In effect, we live in the best of all possible worlds given the resources and technology available to us and the distribution of income.

What set of assumptions produces such an extraordinary outcome, so at variance with our common experience of the real world? First, it is assumed that the economy is atomistic, consisting both of isolated individuals who maximize their satisfaction or utility and of small enterprises that maximize profits. The individuals who take decisions, whether as consumers or producers, are removed from their social context, depersonalized and dehumanized. Each person is an island unto himself or herself and each atom in the system, an "agent" that enters into contracts with other agents and quietly goes about its life maximizing utility or profits as appropriate.

The classical economists of the nineteenth century, and those inspired by them, had a rather different vision of the economy. In their view it was peopled by entrepreneurs, robber barons and distinct social classes with conflicting interests. As David Ricardo (1817), in the first sentence of the preface to his *Principles of Political Economy and Taxation*, asserts, "The produce of the earth—all that is derived from its surface by the united application of labour, machinery, and capital—is divided among three classes of the community, namely, the proprietor of the land, the owner of the stock or capital necessary for its cultivation, and the labourers by whose industry it is cultivated." In Karl Marx's (1867) *Das Kapital,* class conflict played a major role in determining the behavior of the system. The hero of Joseph Schumpeter's (1926) *Theory of Economic Development* is the entrepreneur; recall the "creative destruction" that results from his activities. These visions of the economy, although they vary substantially, share one thing in common: They all place conflict and friction over issues of distribution and justice at the center of their analysis. The anemic agent of late-twentieth-century economics inhabits a different planet from his earlier flesh-and-blood counterparts.

The second core assumption is that economic and political power are largely absent and that something approximating perfect competition prevails. The agents are not only bloodless and depersonalized, they are also powerless. Social groups—be they organized classes in conflict with one another, trade associations, labor unions, lobbyists, special-interest groups, or the military-industrial complex—are of no economic significance. There is no monopoly (single seller) or monopsony (single buyer) and, hence, no super-normal profits, unearned rents, or exploitation. Businesses earn only normal profits (i.e., the opportunity cost of capital). Power as such does not enter into the analysis except perhaps as an afterthought. The assumption of perfect competition ensures that each agent in isolation has no influence on anything that matters. This assumption also makes it possible to conclude that the whole (the overall outcome of the economic system) is no more and no less than the sum of its parts (either a vector of outcomes or the aggregation of decisions by an infinitely large number of agents).

Competition, in this view is seen as a state of affairs, as a general competitive equilibrium established instantaneously by the great auctioneer in the sky. The alternative view, going back to Adam Smith's (1776) *The Wealth of Nations,* regards competition as a process, a mechanism that creates rather than merely equilibrates markets; that probes, seeks, and discovers economic opportunities; that widens horizons; that promotes specialization and a greater division of labor. From this latter perspective, competition is as much concerned with growth and change as with the efficient allocation of given resource and stocks of commodities.[1] Indeed, in Schumpeter's formulation of "monopolistic competition," a market structure that is inefficient from an allocative standpoint—namely, monopoly—plays a critical role in promoting innovation and change and, hence, in overcoming inertia in the economic system. Competition as process rather than as state of affairs also informs Albert Hirschman's (1970) book, *Exit, Voice and Loyalty: Responses to Decline in Firms, Organizations, and States.* This alternative view of competition as process implicitly confronts friction in two senses: as conflict among firms to appropriate profits and as the resistance encountered when entrepreneurs, firms, workers, and consumers rub up against one another in the pursuit of individual or collective interests.

The third core assumption of late twentieth century economics concerns information and expectations. All agents are equally well informed; none are ignorant. It is thus impossible for one agent to take advantage of another by exploiting his lack of information. In a world such as this, price is the only information an agent requires from the market in order to make an efficient decision. In this world no diner needs to consult a sommelier about the characteristics of a wine on a restaurant wine list; no patient need consult a neighbor about the competence of the local dentist; no consumer is mystified by the intricacies of video recorders, computer software packages, automobile fuel injection systems, or the nutritional chemistry of products on sale in a nearby supermarket. All of us are assumed to carry in our heads a comprehensive and up-to-date encyclopedia of the world's knowledge. Indeed, knowledge is a free good and the acquisition of knowledge is costless.

Given the absence of friction, the absence of obstacles to the spread of information, there is little to prevent people from searching thoroughly for the best buy, the best job, the best holiday, and so on. In short, there is little to prevent people—all people, all the time, everywhere—from reaching an optimal position. In this sense, economics tends to suggest that whatever is, is right. In fairness to the conventional view, however, it should be added that late twentieth century economics does allow for the possibility that personal optima may not result in a social optimum. There may be market failures. Nonetheless, the strong commitment to utilitarianism at the level of the individual leads inexorably to the conclusion that whatever the market produces is likely to be best for society as a whole.

When agents look to the future—an enterprise investing in fixed capital, a family planning for retirement—there is of course incomplete information. No one can know exactly what the future will hold. It is commonly assumed by modern economists, however, that the full array of possible events is known and that the probability of each event occurring is also known. In this way, ignorance about the future is reduced to a problem of risk assessment in which the probability calculus can be applied with precision. That is, an uncertain future beset by risks is transformed into a certainty equivalent and the analysis can then proceed as if information were perfect.

The alternative approach, advanced by J. M. Keynes (1936) in his *General Theory of Employment, Interest and Money*, is to postulate generalized uncertainty, whereby anything can happen. The world is full of surprises and events are unpredictable. In such a world, no probability (not even a subjective probability) can be attached to uncertain future events. Rational, optimizing calculation breaks down; agents may have to resort to satisficing,[2] rule-of-thumb, or perhaps less minimizing behavior given arbitrary, fixed output targets; and entrepreneurs may be reduced to relying on what Keynes called "animal spirits."

A common view today is that we not only are able to reduce possible future events to a certainty equivalent but we also fully understand the implications of each event. The encyclopedic knowledge each agent carries around in his head includes complete knowledge of how the economy works, how other agents (including government) will respond to events, and the consequences of the actions of others as they affect the agent concerned. It is thus possible to form fully consistent, rational expectations about the future.[3] In a world such as this, market processes cannot result in crises, panics, waves of destabilizing speculation, or endogenously generated business cycles.[4] Even countervailing government policies are ineffective (and of course unnecessary) because private agents can accurately anticipate them, discount their consequences, and take appropriate offsetting actions of their own. Worse, government action is often seen as the cause of economic instability rather than as a remedy for it.

The fourth core assumption has to do with the treatment of time. Standard late-twentieth-century economics is essentially timeless; little interest is shown in history either as a source of "puzzles" for theorists to solve or as empirical material that can shed light on the validity of theories. As Robert Solow has commented, "Economic theory learns nothing from economic history, and economic history is as much corrupted as enriched by economic theory."[5] Much effort is devoted to an exploration of the characteristics of static, timeless equilibria—either for the economy as a whole (general equilibrium) or for a single market (partial equilibrium). And even when theorists introduce the time dimension, they often do so in a highly formal way merely by adding subscripts of t or $(t+1)$ to variables in an equation or by collapsing the future into the present by assuming that it is possible to enter into contracts in forward markets. An alternative is to conceive of the world as being in dynamic disequilibrium with indeterminate solutions. Such a world is characterized by open-ended outcomes of economic processes in real time. Processes that are influenced in part by expectations, speculations, and anticipations. In much current economic thinking, however, real time is ignored; the irreversibility of history is overlooked, and mathematical or logical time is treated as reversible.[6] A price rise differs from a price fall only by the direction of change; an increase in unemployment differs from an increase in employment only by the sign of the change variable. Investment (i.e., a change in the stock of capital), presents more of a problem, since one can add to the stock of durable assets more quickly than one can reduce the stock of fixed assets, short of scrapping them. More important, technical change is often ignored or treated in a mechanical way as occurring exogenously, randomly, without bias, and as being easily spread.

An alternative approach has recently attracted considerable attention. Known as "path dependence," this approach puts historical time at the center of the explanation of economic phenomena.[7] Analysts of this persuasion argue in effect that one's destination depends on the point of departure and on closely related events along the way. Time is irreversible; one can't go home again. Once embarked upon a particular path, once a particular innovation, say, has been introduced, the subsequent trajectory is largely determined. Minor improvements of the original innovation then, will help to entrench it within the economic system, and subsequent discoveries will be ignored—those innovations will not be introduced—even if they are superior to the original. Once we have chosen a particular color television system or computer hardware technology, pervasive fixed costs, increasing returns, and cumulative processes will ensure that the initial path is followed to its conclusion even if alternative paths would have permitted a faster or cheaper voyage. Similarly, having located the U.S. manufacturing belt in the northeastern region and the eastern part of the Midwest, manufacturing continued to be concentrated in these areas despite the fact that the economic center of gravity had moved to other parts of the country.[8] Inertia in the form of sunk costs in transport, economies of scale, and externalities kept manufacturing where it was. This point is ignored in much contemporary thinking because history and sunk costs are disregarded: Economists are taught that "bygones are bygones."

Few if any economists would admit that they accept the four core assumptions just enumerated and the frictionless economy to which they give rise. The picture I have drawn here is a distortion, a caricature of what most economists believe. The hope, however, is that by exaggerating some of the features of late-twentieth-century economics, I can reveal and highlight part of its true nature. Thus I am not being disingenuous by revealing the essence of modern economics through presentation of a caricature; rather, I am trying to break through the wall of qualifications that surrounds and protects the most widely accepted economics paradigm. No well-trained economist would deny the obvious claim that friction and inertia are absent from the economy; but in the picture presented to university students in elementary textbooks written by the same well-trained economists, friction is conspicuous by its absence. Indeed, students are presented with a picture that is close to the caricature I have drawn.

The Struggle to Incorporate Friction into the Core Assumptions

Economics is in a bit of a quandary from which it is struggling to escape. On the one hand, theories built upon the core assumptions—concerning the atomism of economics, a high degree of competition, perfect information and foresight, and rapidly adjusting prices—yield results that are highly unrealistic, have few policy applications, and open economists to the criticism of inhabiting an ivory tower. On the other hand, theories built on alternative assumptions—sticky prices, animal spirits, generalized uncertainty—are rejected for being ad hoc and arbitrary and for having weak microfoundations. Friction, so to speak, is introduced into the system, and greater realism is thereby attained, but only by ignoring most of the teachings of standard price theory.

Ever since the publication of John Hicks's (1965) *Capital and Growth,* economists have understood how fixed, sticky, or slowly reacting prices could generate instability and a failure of markets to clear. Sticky wages in the labor market, for example, could account for the phenomenon of unemployment. In a "fix-price" world, markets usually adjust to a disturbance by varying quantity rather than price, resulting in the possibility of persistent disequilibrium, oscillation of production and incomes, and, in general, a suboptimal performance of the economy. It is easy to imagine how such outcomes could occur. Assume, for instance, that a low-cost building material, substituting for steel, becomes available on the market.[9] The steel makers notice a falling off of demand, but they do not know whether it is temporary or permanent, a random fluctuation that will quickly be reversed or a structural change in their market. Consequently, their initial reaction may well be to wait and see, to do nothing. Uncertainty results in inertia. The price of steel remains unaltered, production remains constant, and the fall in demand is reflected entirely in a rapid buildup of unwanted stocks. This situation triggers a response in the next period, but the response in a fix-price world is to curtail production sharply in an attempt to bring the level of stocks back to its "normal" relationship to sales. Prices continue to remain unaltered, and the adjustment to the change in demand is now carried entirely by a reduction in the volume of output. Lower output, in turn, is likely to be accompanied by a reduction in employment, not by lower wages. If the steel industry is large, the effects of a contraction of output and employment could spread to the rest of the economy. No doubt prices and wages would begin to adjust after a lag, but by then the damage would have been done and the actual results of the operation of the market mechanism would be very different from those implicit in a model of a frictionless economy.[10]

Granted that sticky prices can help to account for much observed behavior, we are left with the puzzle of what accounts for the sticky prices. One approach emphasizes "transaction costs," in other words, the fact that to create markets in the first place, to enter into binding contracts, to enforce agreements in the courts, and to engage in transactions is to entail costs.[11] Trading is costly; information is incomplete and expensive to obtain; the market system is not a free good; there is friction in the system. Let us consider the implications of this observation, first for commodity markets, then for the labor market, and finally for the capital market.

Price rigidities in commodity markets have been explained in part by "menu costs"—in literal terms, the costs to a restaurant of printing a new menu or the costs to a mail order company of publishing a new catalog.[12] Changing prices entails both real resource costs and psychological costs: the resources needed to set new prices, the effort required to inform customers, the irritation to customers caused by price changes, and even the mental exertions of managers when determining profit maximizing prices. These costs introduce inertia into the price formation process and inhibit instantaneous market adjustment to disturbances. Furthermore, price rigidity is increased if to menu costs one adds complementarities and the resulting possibility of "coordination failure."[13]

Imagine a situation in which it is advantageous for an agent to incur menu costs and change his price only if other agents also do so.[14] This complementarity could arise if firms in the same industry are engaged in monopolistic competition, such that the action of one firm has economic implications for the other (few) firms. In such circumstances, all firms would gain if prices were raised, but no firm acting on its own would raise its price for fear of losing custom. Equally, complementarity could arise between firms supplying inputs and those producing a final product. For example, it might benefit a supplier to reduce the price of his intermediate product only if he can be confident that the buying firm would pass on the lower costs in the form of a lower price of the final product, thereby increasing demand and sales for both final and intermediate products.

Regarding the labor market, it has long been recognized that variations in the level of wages frequently have not acted as a market equilibrating mechanism. The labor market often fails to "clear" in the sense that at the going market wage there are more people seeking work than employers seeking workers. In other words, wages tend to be sticky; friction in the market prevents rapid adjustment to changes in economic circumstances. One explanation put forward to account for this phenomenon is that government policy intended to benefit workers has reduced the efficiency of the labor market and consequently harmed the interests of the working class. In part, it is claimed, the reason is that governments have encouraged or at least allowed the formation of monopolistic trade unions. Collective bargaining by powerful unions has pushed wages above the market clearing level, thereby forcing some workers into either unemployment or poverty, i.e., into residual, informal sectors where wages are flexible. More important, generous unemployment compensation, payable over a long period, combined with social safety nets designed to keep people from falling into poverty, has allegedly created a set of conditions that enables workers to engage in lengthy job searches but provides little incentive to unemployed people to accept offers of employment that fail to satisfy their aspirations regarding income, location of employment, and job characteristics. High unemployment and poverty, according to this view, are products of the welfare state.

Government interference in the labor market is contrasted with the laissez-faire that prevailed in the nineteenth century. During that period, it is said, labor markets functioned efficiently: Wages were flexible, contracts were short (perhaps no longer than a day), and turnover was high. Labor markets approximated the "spot" markets of economic theory. Recent research, however, has shown that this view of nineteenth- and early-twentieth-century labor markets is erroneous in all respects: Most workers were not hired on a "casual" basis, contracts were not short, and wages were not particularly flexible.[15] Moreover, as the twentieth century proceeded, employers created internal labor markets within their firms, leading to sticky wages and reduced labor turnover. Yet the development of internal labor markets had little to do with government interference. We therefore must seek elsewhere for an adequate explanation of sticky wages.

The so-called efficiency wage hypothesis has been advanced to explain why real wages (i.e., nominal wages adjusted for inflation) do not fall in the face of a decline in demand. Originally intended to explain "surplus" labor in developing countries, the efficiency wage hypothesis is now also applied in developed countries. The original idea postulated a causal chain linking wages to food consumption, nutrition, productivity, and unit labor costs.[16] Employers paid more than the market clearing wage because a higher wage resulted in improved nutrition, enhanced productivity, lower unit labor costs, and higher profits. More recent versions of the hypothesis maintain the link between wages and productivity but place much less emphasis on nutrition.[17]

In these models employers pay workers more than a market clearing wage in order to reduce labor turnover. Stability of the labor force (low turnover) may reduce unit labor costs because of learning effects associated with job experience in a specific firm. Alternatively, a reduction in real wages could result in a lower average quality of the labor force. Lower wages (because of "adverse selection") would induce the best workers to quit first, thus lowering average labor productivity and profits. A final possibility is that employers may pay high wages in order to reduce the cost of monitoring labor or to elicit greater effort from the work force. After all, high-wage workers are less likely to "shirk" on the job because they have more to lose if caught and fired. Labor effort and output per worker-hour therefore rise with the wage rate, and costs of supervising labor vary inversely with the wage rate, such that, over the relevant range, higher wages lead to higher profits.

However they are caused, rigidities in the labor market, combined with price stickiness in commodity markets, can go quite far in explaining the observed behavior of actual economies.[18] Menu costs, other transaction costs, and coordination failure introduce inertia into commodity markets and inhibit a downward adjustment in prices when demand falls. Moreover, sticky real wages inhibit adjustment in the labor market, preventing unit labor costs from falling and providing an additional reason for firms not to reduce prices when demand declines. The two phenomena interact, reinforce one another, and prevent the economy from achieving equilibrium. In fact, disequilibrium can persist more or less indefinitely.

Friction in the capital market compounds the problem. The four core assumptions of late twentieth century economics are intended to help us understand how a market economy functions—in short, how capitalism works. Yet there is considerable recognition among economists, that the capital market—perhaps the center of gravity of a capitalist system—functions less like the market of price theory than any other market. Paradoxically, then, the theory of capitalist markets does not explain the functioning of capital markets. Indeed, capital markets by their very nature are in disequilibrium; the market for capital never clears.[19]

The reason is that capital markets are riddled with risk, uncertainty, informational imperfections, and expectations that after the fact turn out to be highly mistaken. The price of capital (e.g., an interest rate), does not convey enough information to allow the market to function efficiently. The seller (lender) often needs personal information about the buyer (borrower) in order to assess the likelihood of default on the loan and the consequent loss of part or all of the lender's capital. Impersonal, atomistic transactions among faceless agents with complete information are the exception rather than the rule. The interest rate in a market for loanable funds never equilibrates supply and demand. At times, as in a deep depression, there is an excess of loanable funds. More typically, there are more potential borrowers at the going rate of interest than there are funds to lend. Credit is therefore rationed. What, then, accounts for these characteristics of the capital market?

First, information between lender and borrower is necessarily asymmetrical. The lender, unable to read the mind of the borrower, cannot fully assess the likelihood of default.[20] He cannot be sure, for example, that the borrower will not abscond with the funds or vanish when the time comes to repay. The risk attached to a loan cannot be known; there can be no certainty equivalence of a loan portfolio. Second, the lender, though knowing that the likelihood of default depends in part on alternative actions open to the borrower, nonetheless has only a limited ability to control those actions or monitor them. The lender, in other words, has little alternative but to rely on the good sense of the borrower. Third, a loan differs from most other market transactions in that it is not completed instantaneously but extends over time, in some cases over a prolonged period of time. A loan consists of an exchange of money "today" in return for a promise to repay the money in the future with interest. Once the future is brought into the transaction, generalized uncertainty becomes relevant and the certainties of a spot transaction disappear.

As a result, the loanable funds market differs from commodity markets by the presence of nonprice features: collateral (to reduce the consequences of default), loan conditions other than interest rates (to influence the behavior of borrowers and reduce the risk of default), and credit rationing (to screen out less creditworthy borrowers). Unlike the market for widgets, the allocation of loans is not left to the forces of supply and demand. He who offers to pay the highest interest rate does not necessarily get a loan. Seen from the perspective of a bank or building society, the highest competitive price—the market clearing rate of interest—does not necessarily represent the profit maximizing price for lenders in the capital market.

Thus there are good grounds for believing that actual economies do not function in the way suggested by theories built on the four core assumptions. Friction is pervasive in the economic system. As a result, commodity prices can be sticky both upward and downward, whereas wage rates tend to be sticky downward and interest rates sticky upward. Rarely if ever do markets clear, and they certainly do not clear simultaneously as general equilibrium theory would have us believe. The economy is in a permanent but changing state of disequilibrium.

Friction as Obstacle versus Friction as Binding Agent

The present chapter may seem to have implied thus far that friction and inertia, though perhaps inevitable, are undesirable features of all real economies. The less we have of them the better. But is this necessarily true?

At one level it seems fairly obvious that we would want prices to be accurate signals of costs and benefits. Anything that prevents prices from performing their signaling function, causes them to be sticky, and introduces inertia is likely to result in inefficiency in the way resources are used as well as in a failure to use all resources fully. Similarly, anything that prevents an economy from responding to changes in relative prices, such as the specificity of physical capital, or that reduces the elasticity of supply and introduces friction, is likely to result in sluggish adjustment to external disturbances and a lower average rate of growth. Those economies that can more readily transform swords into ploughshares or vineyards into textile mills will tend to be better off than those that cannot. Friction and inertia evidently can act as obstacles to smooth adjustment and steady progress.

At another level, however, it is not so obvious that less inertia, more price flexibility, a faster rate of adjustment would necessarily be desirable. In some circumstances, friction in the sense of resistance to relative motion may act as a binding agent, a stabilizing device. This is particularly true of sticky prices. Consider the economy of the United States, the world's largest market economy.

It is generally agreed that in the United States wage and price flexibility have been lower in the period since the end of the World War II than in the nineteenth century and the first forty-one years of the twentieth century. It is also agreed that output and incomes have been less volatile in the postwar period than they were earlier. Could there be a causal connection between the two contentions? Could price stickiness result in greater rather than less stability of output and employment? It is possible to argue that greater price stickiness and greater output stability were both caused by a common third factor—namely, fewer and less severe exogenous shocks.[21] This view, however, leaves open the question as to whether "shocks" can properly be regarded as truly "exogenous," like a meteor from outer space, or whether they should be seen as endogenous products of the economic system. And if they are endogenous, we need to know why in the postwar era the frequency and severity of shocks suddenly declined.

One possibility is that the rise of a large nonmarket sector—big government—has introduced considerable inertia into the economic system and thereby acted as a stabilizing device. The sluggish response of government to price signals, the cumbersomeness of the budgetary process, and the absence of a short-run, cost-minimizing ethos in government all help to prevent market-induced downward spirals from gathering momentum. Leviathan has become a huge shock absorber, and the friction embodied in governmental institutions may account in part for the reduction in downward instability in the years since World War II. In other words, it may be that what has changed is not the incidence of shocks but the ability of the system to cope with shocks.

Economists in the Keynesian tradition have long argued that expectations are a causal link between sticky prices and stable output and incomes. In this view, information is inevitably incomplete and uncertainty is rife. If prices in general fall, perhaps because of insufficient demand, producers may expect them to continue to fall and consequently cut back on production, thereby further reducing aggregate demand and creating conditions for a second round of price and output reductions. Price flexibility, in other words, may lead to destabilizing expectations rather than to stabilizing adjustments in output. Greater price inertia, on the other hand, might actually help to reduce fluctuations of the economy.[22]

The U.S. experience of the Great Depression of the 1930s is also suggestive. Between 1929 and 1932, output in the United States declined by about a third, yet prices during that period declined by 9 percent a year. It is hard to believe that output fell so steeply because prices did not adjust quickly enough to excess supply or that the depression would have been shorter and less severe if prices had been even more flexible. On the contrary, if wages and commodity prices had fallen further and faster, the depression would probably have been even worse.[23] Had there been more inertia and friction in the economy, the effects of the depression might have been attenuated.

Finally, let us consider more recent events. In the 1980s the global economy became much more closely integrated, capital markets were liberalized, and exchange rates were allowed to fluctuate freely in response to the forces of supply and demand. The consequences have been increased volatility of world output, a reduction in the average rate of growth of the world economy, erratic variations in real rates of interest, and sometimes violent fluctuations in exchange rates. Looking back, we find that the institutionalized inertia of fixed exchange rates under the Bretton Woods system coincided with a golden age of prosperity and stability.

Of course, these historical anecdotes do not prove that friction is an inevitable as well as sometimes desirable feature of an economy, but they do give one pause for thought. The frictionless economy associated with the four core assumptions of late-twentieth-century economics is not just an impossible world; it is probably also a highly unstable world. Economies need brakes when they head for a crash, a sea anchor when riding out a storm; and people, when they go about their daily business of earning a living, need not just explicit contracts but other people they can trust and institutions (including government) on which they can rely.[24] In the broadest sense, then, friction acts as the glue that holds the economy together, as a binding agent that prevents the separate parts of the economy from flying apart. So, although friction is inevitable and awkward, it is also essential.

This can be put another way. The high cost of transactions frequently leads to exchanges occurring in nonmarket settings, within firms, government bureaucracies, universities, and other institutions. These nonmarket relationships among people, when first introduced, are likely to be more efficient than market transactions, since the absence of markets is often an indication that the high cost of transactions makes exchanges mediated through the market unprofitable. At the same time, nonmarket transactions are sticky; they exhibit a high degree of stability; they are full of friction. In short, friction, stability, and efficiency are bound together in these instances.

Such instances are numerous. Indeed, they are as numerous as the institutions one finds in any economy. In a frictionless world there would be no need for economic institutions of any sort: Each atomistic individual would constitute a one-person enterprise and enter into contracts with as many other one-person enterprises as necessary. We would all be self-employed; each person would be a so-called profit center. Frictions, transaction costs, and absent and incomplete markets make such an economic organization of society impossible. Instead, we have institutions that regularize relationships, internalize what might otherwise be market relationships, and, where there are markets, temper the free play of market forces. In so doing they introduce both stability and efficiency into our economic lives.

7

Essential Friction: Error-Control in Organizational Behavior

Gene I. Rochlin

Introduction

In the physical world, interaction without friction is nearly inconceivable. The most elaborate mechanisms would be required to pick up a bottle, set a table, drive a car, or walk to the door. Objects set in motion, whether pencils, mice, feet, automobiles, or trains, would slide freely until arrested or deflected by collision with another object. Without the damping effect of friction, we would live in an impossibly kinetic world in which the consequences of every action would persist and multiply to the point of insanity.

Friction is a mediator of sanity in the social world as well. In the realm of the scientific and technical, the source of friction is a complex of micro-mechanical actions that convert the energy of motion to heat, and the net gross momentum to a series of small internal vibrations. In the realm of the social and political, morals, ethics, knowledge, history, and memory may all serve as the sources of "social friction," by which gross motions are damped, impetuous ones slowed, and historical ones absorbed. Such friction is essential to prevent the persistence and multiplication of social and political movements once their driving force is removed.

In a complex, highly organized, very technical modern society, the role of friction in organizations is also essential, though at best very poorly understood. Whether they be formal or informal, localized or international, organizations of production, of regulation, of administration, or of governance, these elaborate networks and structures are often the primary means of implementation and action. They also contain within them currents of historical momentum, loci of impetuous action, and sources and modes of error from which society should be protected. Social organizations can present very real dangers to the society they are designed to serve if they do not contain effective mechanisms to provide frictional balancing and damping for impulses and drives that would otherwise continue unchecked.

The essentiality of "organizational friction" is therefore more than a redefinition of the obvious need for internal mechanisms of control. Does a corporation lobby for permission to continue manufacturing traditional gas-guzzlers rather than switch to more fuel-efficient cars? Has a plant manager decided that hour-long lunches are a waste of time or that all employees must participate in EST seminars? Have the police decided to ignore *habeas corpus*, or have airport guards decided to carry submachine guns? Has a legislator proposed to attach to a budget authorization a prohibition on abortion or an amendment to allow the executive to red-line the budget? Has some middle-level bureaucrat decided to replace the computers in air traffic control with newer models all at once or that nuclear power plant operators must have paramilitary rank or wear fancy uniforms?

However ridiculous these examples sound, all of them have been suggested at one (recent) time or another. In each case, there were organizational or institutional mechanisms by which the impulsive energy could be dissipated through argumentation, discussion, and other, more bureaucratic means. As in the physical world, those with the most energy were best able to overcome the friction; impulses that proved to be momentary, erratic, or poorly supported tended instead to damp out.

What prevents notions of great transient impulse but with little or contro-verted sustaining support from setting into motion large social forces is in-deed organizational friction. Of course, such friction also tends to slow down the decision process, even when the decisions are correct or proper. A classic example was provided by an anonymous American bureaucrat in a recent editorial, while complaining about an organization whose direction and mandate had been crippled by the Bush administration: "The best and brightest at my agency and others dutifully exercise caution in substantive matters, avoid action and continually seek another clearance, another signa-ture, another authorization until someone just finally says no" (Anonymous, 1992). Too often, this is mislabeled as "inertia," particularly when applied to formal bureaus. The distinction is critical: Inertia is a measure of the force that must be applied to get a bureau or other organization to initiate movement or change its direction, whereas friction is a measure of the energy required to keep the bureau moving or, inversely, the rate at which movement will decay once the energy for motion is removed.

There certainly are inertial laws at work. An organization at rest tends to remain at rest; without friction, an organization in motion tends to continue in motion. When that organization is performing the wrong task, or performing it badly, it may well continue to do so. And though it is true that an erring organization can be corrected, punished, or at least threatened, this fact is small solace when the task the organization is called upon to perform is critical for social well-being or public safety. Friction may therefore be an antidote to the inertia of error; mechanisms for debate, for review, and for oversight of prospective decisions and important actions give an organization embarking on an incorrect or errant course the opportunity to review and, possibly, to recover before major problems occur.

Most traditional analysis has tended to focus on the efficiency with which an organization carries out its tasks rather than on the nature of outcomes or the importance of mistakes. The role of friction in minimizing the propagating internal effects and external consequences of individual or organizational error is thus often overlooked and frequently misunderstood—particularly in the case of organizations that could be a source of considerable social risk if their hazards were not controlled, or those whose primary service role is overseeing, regulating, or controlling the risk of inherently hazardous operations (Rochlin, 1992). Those who seek to remove organizational friction in the name of efficiency or productivity often do not perceive that they may be removing an essential mechanism for social control for the sake of what, in many cases, is more an ideology than a rational plan of action (Perrow, 1986; Rabinbach, 1992).

Taylorism, "Fordism," and the Search for Efficiency

The modern search for organizational efficiency has its roots in the reforms of the "scientific management" movement of the late nineteenth and early twentieth centuries. As with many succeeding reforms, the original purpose was the reduction of social friction—"cures for low productivity, low morale, high error rates, and high wastage of human and material resources" (Merkle, 1980, p. 1).

The early form of industrial organization may fairly be described as a piecemeal aggregation of individual units (Hirschhorn, 1984). In some cases, work might be broken up into individual tasks according to skills and specialization, but it was more common for a single worker to carry one job or task through from beginning to end. As the Industrial Revolution took hold, the nature and definition of "craft" degenerated as the partitioning of labor narrowed the scope and definition of the individual task (Doray, 1988, pp. 34ff.). Increasing specialization following the perfection of mass production began to create a series of problems, social frictions that undercut the greater productive efficiency that had been sought. From the viewpoint of managers seeking control, planning and coordination were inadequate and poorly informed; information, when available, was often unreliable and incomplete; and scheduling was at best difficult (Waring, 1991, p. 10).

Such problems were exacerbated by the creation of large factories or industries that employed large numbers of people to perform tasks by rote on a piecework basis. The simple laborer rather than the skilled worker became the productive core (Doray, 1988, pp. 61ff.). Industrial workers were increasingly disempowered; the more ambitious, motivated, and intelligent ones were quick to note that income and social mobility were increasingly associated with soft jobs such as management and sales rather than with physical work, however skilled. Intelligence and experience were therefore leaving the plant and shop floor even before the first wave of reform. Among the negative side effects was a decreased incentive for workers to identify problems, to take initiatives, or to work any harder than was absolutely required.[1]

It was in this context that Frederick W. Taylor was moved to introduce the principles of what he called "scientific" management, derived largely from the application of production engineering.[2] As later critics have pointed out (Merkle, 1980; Waring, 1991), Taylorism was more of an ideology than a management system in the modern sense. What Taylor himself sought was the "one best way" of reducing conflict, within the plant and without, an attempt not only to reduce, but to minimize social friction in the broadest sense. As Judith Merkle points out (1980, p. 4), Taylor also sought to curb "disorderly" informal modes of organization and to combat what he perceived to be a persistent pattern of "soldiering" and low motivation among workers (Taylor, 1987). Instead of proposing social or factory reform, which would have been critical of management, Taylor suggested that these results could be brought about through a series of mechanical and organizational devices.

Although Taylor had only moderate success in getting his agenda adopted during his lifetime, many of the underlying precepts of Taylorism were to become broadly accepted as part of the growing trend toward rationalization of industrial organization (Simon, 1976). At the turn of the century, the increasing sophistication of process and management led to a systematic effort for greater efficiency, including systematic attempts to further eliminate "wasteful" duplication of skills. In the electrical and chemical industries, the entire means of production was radically revised to coordinate with the shift in emphasis from practical to scientific knowledge (Noble, 1986, p. 16; Rabinbach, 1992, pp. 238ff.). The principles of Taylorism coincided neatly with this agenda, as one of the aims was to remove the control that the craft guilds had heretofore retained through their monopoly on experiential knowledge and expertise. The workplace was to be socially reorganized to separate the design, organization, and management of work from its execution (Taylor, 1987). As craft procedures became increasingly and more formally codified, craftsmen were to be eased out and replaced by less skilled workers performing a limited range of tasks learned by repetition and rote (Doray, 1988, pp. 34ff.).

Taylorist principles were also in evidence when Henry Ford developed the first mechanical assembly line (Nevins, 1954). In the classic Fordist model, individual tasks are sequenced together tightly by mechanical means (the "line"), whose underlying structure and purpose is the increase of productive efficiency. Factories lower costs by reducing "slack" (available but normally unused resources), buffering capacity (resource storage space between or among different tasks to allow for uneven output rates), skill duplication, and the need and means for interworker negotiation. This was to be achieved first by standardizing, and then by tightly coordinating and scheduling individual tasks and suboperations.

The friction so reduced on the production side was, however, purchased at considerable cost to worker morale (Merkle, 1980; Zuboff, 1984). The social mechanisms of interaction that convey pride in work and distribute responsibility for overall output in craft or skilled piecework operations are also sources of friction. By isolating workers in the name of efficiency, the assembly line removes both communication and understanding. Each worker is assigned a task so narrow that accumulation of knowledge seems essentially irrelevant, and is given little information as to the relationship between his or her work and that of others. In such an organization, the individual's domain of perception of the consequences of error are severely restricted; there is minimal knowledge of how error will propagate through the plant, or how to detect and correct for the propagating errors of others.

As will be discussed more fully below, the tightness of process coupling that characterizes the traditional assembly line also creates situations in which new types of "systemic" errors emanating from collective properties, interactions, structure, or design can be generated that arise from, but not because of, individual action. Without informed social interaction, the likelihood of their detection is small. Removal of internal friction therefore threatens to create situations in which uncorrected internal error could bring the whole operation to a halt while consequences and causes are being disentangled.

On the technical side, those who had risen into the "white-collar" technician class from the plant floor began to be displaced by "school-trained" professional engineers, whose understanding of management practices and objectives, coupled with knowledge of the principles and general theory of their field, was considered more important than detailed knowledge of any particular plant or process (Noble, 1977; Rabinbach, 1992). Toward the middle of this century, industrialists and managers who had come up through their own industry from a beginning in small-shop craftsmanship and managerial entrepreneurship—such as Henry Ford and Donald Douglass—became increasingly rare even in management. Preference was increasingly being given to those with professional education and certifiable management skills.

The evolutionary development of large-scale technical-industrial concerns therefore tended to favor those whose education and knowledge of the plant was largely professionalized and formal, further decreasing the status and visibility of those whose knowledge was primarily experiential. As engineers became more professionalized, their familiarity with operational details decreased; as processes became more sophisticated, the range of competence of individual workers was a decreasing fraction of the whole.

This is not to say that the "experts" ceased to exist. Indeed, as plant and process grew more complex, the role of experiential knowledge in maintaining an integrative view of operations tended to increase (Dreyfus and Dreyfus, 1986). In many cases, the role of and respect for those who had managed to acquire and retain experiential knowledge despite the many changes in plant and managerial design became more prominent: "old Mike" on the plant floor, who could tell a boiler was going to break down by the way it sounded; "old Harry" in management, who seemed to be able to smell how a certain change would affect sales; "old Eloise" in accounting, who was the only one who could accurately predict whether operations would be profitable that quarter.

Ironically, although such "experts" played a crucial role in mediating differences and managing complexity, their comparative lack of status, power, and decisionmaking authority made them increasingly vulnerable in the continuing search for formal solutions and methods of achieving efficiency (Doray, 1988; Rabinbach, 1992). The scientific management movement's eventual claim to be able to codify and formalize even experiential knowledge carried with it the implication that "old so-and-so," however quaint, was no longer central to the operation of the plant or firm (Zuboff, 1984). To compensate for the loss of skilled workers, oversight at the level of production or process was increasingly concentrated in foremen and other intermediate managerial classes, under the assumption that their oversight of production and task was sufficient to ensure coordination and integration (Doray, 1988; Waring, 1991). At the same time, the increasing shift in importance of the machinery compared to the workers elevated technical cadres, "whitening" their collars.

As process sophistication grew, so did the importance, power, and status of technicians and engineers—with the highly professionalized, management-oriented cadre of "school trainees" relatively remote from the actuality of plant and procedure gradually displacing the older "shop school" in one industry after another. These specialists were trained to put a premium on efficiency, and to regard the internal social processes of the plant floor primarily as sources of operational friction (Simon, 1976). Moreover, for the engineering profession, operational friction was (and remains) merely another inefficiency to be sought out and extirpated. Although technology is fast reforming the assembly line in many traditional manufacturing industries, the agenda of efficiency—reductionist separation of function—continues in many others (Blau et al., 1976; Zuboff, 1984).

The period of the "old hands" is now remembered with some nostalgia and considerable fondness, but such people were rarely well rewarded either for their value to the firm or for the knowledge they made available to those who sought to codify and formalize it. Nor were the firms that replaced them always able to capture the desired gains in efficiency. With the transition from skilled to standardized labor, more and more oversight was required for process reliability, which increased organizational complexity and the need for management coordination. Despite the loss of traditional measures of productivity, the extensive clerical and managerial staff that were required were now seen as necessary and productive labor rather than as expensive and wasteful consumers of resources (Merkle, 1980, p. 76).

By the 1920s, Taylor's agenda had played itself out, overtaken by other management schemes and methods and by the unionization of labor, which resulted in greater emphasis on cooperation and social rewards (Doray, 1988; Perrow, 1986; Waring, 1991). An increased focus on the role of managers led to increased training and professionalization, especially of middle management; and both trends tended to reduce the managers' autonomy as well as the imposition of arbitrary hierarchical control. During the 1930s, the work of Mayo and Chester Barnard, emphasizing social roles and group norms, completed the agenda of social transformation. Instead of seeing workers as isolated productive units, the organization was now viewed as a primarily cooperative, socially integrated enterprise (Barnard, 1938; Perrow, 1986; Scott, 1981).

Barnard's work pointed to the three main modes of analysis that were to dominate organization theory well into the 1960s (Perrow, 1986): the institutional-bureaucratic approach; models of rational decisionmaking and its limitations, and the human relations school. The underlying premises of moral purpose and social cooperation emphasized the importance and purpose of the organization rather than the individual, undercutting the class-ridden and arbitrary basis of authority assumed by Taylor and other management theorists of the nineteenth century. Managers now sought to legitimate their right to command, and to control, through a variety of social processes such as communication, cultivation of loyalty, worker identification, and so on (Merkle, 1980; Waring, 1991; Zuboff, 1984).

What was to revise and reinvigorate the ideology of quantitative scientific management was the transfer of the techniques and modalities of operations research and systems analysis to the business world, fostered and augmented by the rapid development of computers, at first in the guise of automated data processing equipment. But the new agenda differed in at least one major respect; instead of being viewed as a series of isolated, interacting operations, the firm was now visualized as a single, complex, interrelated pattern of activities, to be analyzed, integrated, and optimized as a whole (Waring, 1991, p. 25).

In retrospect, Taylorism and Fordism constituted a transitional phase from the early phase of industrialization. Whereas the early scientific management theorists and practitioners saw the organization as a means to increase the efficiency of the individual, modern "post-Barnard" analysis tends to focus instead on shaping employee behavior to optimize the performance of the organization (Thompson, 1967). But the tools available for increasing organizational efficiency were at first largely restricted to the social, given the early limits on more quantitative aspects such as data collection and processing and the handling of information flows.

Neo-Taylorism and the Ideal of Frictionless Decisionmaking

Since the 1930s, the emphasis on reducing friction in the shop and on the plant floor has been losing its centrality and glamor as the labor force in industrial countries has gradually shifted from blue to white collar. The general decrease in the fraction of product costs attributable to labor and the growing importance of nonmanufacturing industries, businesses, and other forms of regulatory or administrative organizations in the economy has drawn increasing attention instead to the softer tasks of desk workers. The emphasis in reducing friction has shifted from the assembly line to administrative and managerial efficiency, particularly for the growing number of organizations whose primary "product" or output is essentially social, such as finance, regulation, or coordination. The concept of friction has also shifted. Theorists and analysts are now more likely to be studying decisionmaking, networking, communication, and the management and flow of data and information than such traditional sociological factors as social class, worker identification, and hierarchical relationships.

Much of the new literature on organization theory emphasizes the function of uncertainty in decisionmaking, the role of informal networks and of learning as a goal, the limits of planning and rational action, and the permeability of organizational boundaries. As tasks, goals, structures, and social interactions grew more complex, organizations were increasingly visualized not as self-contained entities but as "natural" or "open" systems (Perrow, 1986; Scott, 1981). Open-system approaches, with their emphasis on the limitations of formal analysis and closed models, now dominate even the rational-analysis school. Indeed, much of the groundbreaking work of Herbert Simon, and of Richard Cyert and James March, was directed primarily at the problem of decisionmaking under uncertainty and with imperfect information (Simon, 1976; Cyert and March, 1963).

But the desire for greater closure did not thereby abate. Much of the literature on management continued to adopt the rational model based on closed-system models, emphasizing the importance of formal planning, goal setting, and control of outcomes rather than that of adaptation and negotiation (Thompson, 1967). Increasing the control and efficiency of use of information therefore remains of central importance to business managers, as it has been since their creation as an identifiable class in the late nineteenth century (Chandler, 1977; Waring, 1991). Until World War II, however, there were few means for systematically addressing either.

The war spurred technical development in a number of critical areas, particularly automation, communication, and information processing, and created a number of new "technological" forms whose intrinsic properties, and risks, not only dominated the next phase of technical industrialization but also created a new category of organizations whose primary function was to control or regulate the technology or its social uses. Perhaps more important for the evolution of business management was the creation of an effective instrument of war that at first escaped much public notice. Lacking photogenic equipment, charismatic heroes, or large and visible pyrotechnics, the twin disciplines of operations research and systems analysis remained generally unknown to the public, even though they would be as responsible for the Allied victory as the more dramatic weapons that brought the fighting to a close (Kittel, 1947).

From the war experience grew a cadre of operations-trained researchers whose interests extended not only to the general theory of systems but also to the new field of cybernetics, with its emphasis on self-regulating "open" systems and quantitative methods. During the 1950s, operations researchers and their descendants began to move into industry and business as advisers and consultants, leading to two currents that have shaped contemporary management science. On the theoretical side, the dominant human relations model of management was to be overtaken by the new systems school in the mid-1960s, with new work such as that of James Thompson (1967) and Daniel Katz and Robert Kahn (1966) complementing the rational-actor models of Simon, March, Cyert, and others. On the practical side, as the systems perspective moved into the business schools, managers skilled in quantitative analysis and systems thinking became permanent members of their organizations, aided by the expansion of computer power and availability.

By the 1970s, mathematical and statistical tools were commonly available, and many of the new class of "scientific managers" trained on and completely familiar with the increasingly powerful computerized tools and methods were using their quantitative skills to move up the management ladder.[3] The advent of the personal computer in the 1980s, and the accompanying dramatic increase in the capabilities and power of all forms of computing equipment, allowed them to bring their tools as well as their techniques into common use at all levels of the firm, from the "shop floor" (where there is one) to the office of the CEO (Zuboff, 1984).

In parallel with these trends, economies of size and scale in some industries (e.g., large chemical plants) led to larger and more automated machinery, larger batches, and quicker processes, while a whole new category of potential products ranging from pesticides to plastics used larger quantities of more toxic and hazardous chemicals. For a variety of newly emerging technologies, such as higher-performance aircraft, nuclear power plants, high-performance military systems, extensive and tightly coupled power grids, and an increasing range of multiproduct, multifunction chemical factories and refineries, the processes themselves were considerably more complex technically than their earlier counterparts, requiring a reorganization of the work into a correspondingly complex structure (Hughes, 1983; Roberts and Gargano, 1990; Weick, 1979, 1987). But as is well known in the organizational literature, effective management under these conditions fosters, and sometimes requires, the construction of informal networks to bypass hierarchical bottlenecks (Scott, 1981).

The regulatory and managerial conditions for operating these large, complex technical systems safely and reliably are also considerable, thus further increasing both the size of the operating organization and the complexity of its decision processes (La Porte, 1982; Perrow, 1986; Thompson, 1967). Where public safety is involved, there may be requirements to submit or report to external regulatory and oversight organizations. This increases decisionmaking and operational friction both through the regulatory process itself and through the requirements that may be imposed on the operating organization (La Porte, 1988). For some highly regulated industries, whole departments have been created whose sole function is to oversee and report on regulatory requirements and produce the vast amount of paperwork necessary to verify compliance. At the same time, the usual pressures for reducing friction—the search for efficiency, reliability, and control—remain present.

As their earlier counterparts were, the managers and administrators of these and similar organizations are responsive to the ideas and ideals of scientific management, which increasingly tends to focus on social organization and interpersonal relationships. But where technology is central to the organization, or to the task it is assigned to perform, the requirements of management extend past the usual array of socially oriented tools (Thompson, 1967; von Glinow and Mohrman, 1990). Moreover, the managers of technically oriented organizations are more likely than most to be fully familiar with and enthusiastic about the use of computers and other forms of automated operational, informational, and data-processing equipment. It is therefore not entirely surprising that designers and managers of many of these organizations have turned to computers and computer systems in search of new methods and new techniques for increasing efficiency, safety, and reliability without accompanying increases in social complexity and social friction (Blau et al., 1976; Danziger, 1986; Weinberg, 1990).

Many elements of the resulting agenda are decidedly neo-Taylorist in their aspirations and implementation. Just as Taylorism declassed a generation of craftworkers, plant automation has transformed the tasks of many workers from active involvement to managers and overseers of the automated equipment that does the actual work (Beniger, 1986; Hirschhorn, 1984; Noble, 1986). This circumstance in itself is not necessarily a shortcoming; as Shoshana Zuboff (1984) has pointed out, the "textualization" of the work environment requires increased knowledge of operations, which in many cases has given workers a greater sense of control over their own work. But even in her own extensive case studies, she noted many other instances where the transfer from hands-on to remote operation meant a relative insensitivity to, and sometimes neglect of, process demands and requirements.

More to the point, the attitudes of many of those responsible for the design and implementation of the new systems have decidedly Taylorist resonances—right down to the scientistic belief that there is "one best way" to perform any task, including a managerial one. Even more striking is the return to the search for central control (Beniger, 1986). In her interviewing, Zuboff (1984, pp. 246ff.) noted several instances where the manager's view of the intended result was centralization of control in order to reduce uncertainty and allow for syncretic planning of all operations. As she reported, a survey of plant managers conducted by the Honeywell Corporation found that "almost without exception, the "technology ideal" reported by plant managers was having one screen in their office from which they could operate the entire plant" (p. 283). In another cited study, Robert Howard (1985) finds that information systems "are indeed being used to reproduce the logic of scientific management—top down control, centralization of knowledge, de-skilling—more comprehensively than ever before."

For most organizations—bakeries, food processing plants, tool makers, stockbrokers, even automobile production—direct social consequences are largely limited to the workforce and locale that are involved. Consumers, and the public at large, bear the burden only through the indirect costs of worker alienation, poor product quality, and individual irritation, and through their own inability to vote on or judge whether the presumptive economic benefits returned are worth the entailed costs. But there are a growing number of organizations in modern, highly technological societies for which the general public bears a large part of the direct cost and will directly feel the effects of errors or other operational failures. The use of automation and computerization to reduce internal friction and increase productivity may in such cases cause considerable harm not localized to the firm or its labor force.[4] So the wisdom, indeed the morality, of "experimentation" in such organizations is far more questionable.

Complexity as A Source of Error

The emergence of a new class of industrial technologies and systems with new levels of technical complexity also carries with it entirely new modalities of failure and risk. Perhaps the clearest historical example is the commercialization of the air transport, based on a technology stunningly unforgiving of errors, whether in operation or in application of technique. For a variety of new industries, technological and managerial complexity is relatively high, and intrinsic coupling relatively tight. And many such industries have a history of failure and accident in their developmental stages, while experiential knowledge was being accumulated. But in the past, societies had relatively low expectations of the performance of new and "experimental" technologies. They could afford to be tolerant of errors so long as adverse impacts were localized and risks were assumed voluntarily. For example, it was only toward the end of the 1930s that commercial airlines reorganized their operations and solicited external regulation to provide the kind of external reassurance that would draw a wide base of passengers.

Widespread commercialization, however, meant a decreased public tolerance for errors and failures. As technologies such as airlines, electrical grids, or nuclear power moved from small to large scale after World War II, potential consequences became not only more widespread but in some ways more likely (Rasmussen and Batstone, 1989). Large commercial nuclear reactors could not self-cool or self-recover in an emergency, requiring the provision of external emergency cooling systems; and the design choice was such that the time available for action in a serious event chain was measured in minutes. Higher-altitude, faster jetliners were less forgiving of design, maintenance, or pilot errors, while an increasing public market made the consequences of each accident far more socially severe.

With the growth in communications and media (particularly TV), neither the outcomes nor the potential for harm could be localized or minimized. At the same time, public tolerance for physical or environmental harm was steadily decreasing, a trend that accelerated with the advent of the environmental movement in the 1960s and 1970s. As a result, steadily increasing pressure was brought to bear on those responsible for the reliability and safety of systems operation.[5]

When errors did occur in such large, complex systems, particularly errors that involved measurable public harm, blame had to be assigned. As a rule, it was almost always the operator who bore it—even when the system design was such that an operator, however well trained, would have neither an adequate framework in which to make an informed decision nor the time to reason through the alternatives (Rochlin, 1992). This assignment of responsibility was most often made on the grounds that the operator "could have" made the right decision—even if the making of that correct decision in the given situation, and in the time available, would have been more a matter of intuition or chance than of rational analysis and procedural action (Perrow, 1984; Weick, 1987)

Given this turn of events, there were, and are, two general paths to reducing the chance of error and minimizing the risk of system failure. The first is to redesign the operator-technology interface to provide more internal discussion, more internal error checking and redundancy, and more processes for internal oversight, collaborative decisionmaking and review— all providing what I have characterized in this chapter as "essential" friction. Greater assurance of system reliability and safety would thereby be gained at the cost of decreasing the rapidity of the decision process and increasing the number of units involved.

Unfortunately, this alternative is more often than not translated by organizational and system designers as an inefficient, even "wasteful" use of organizational resources. More commonly, designers try the other, neo-Taylorist approach, searching for the provision of means by which operator decisions can be supported but also, in many cases, replaced by fully automatic equipment (Brissy, 1989). Such equipment, and the plant and process models upon which it is based, is often held up as an ideal goal to be sought, capable of responding much more quickly in an emergency and far less subject to the vagaries of human reasoning and judgment (Ralph, 1988). Indeed, computer-controlled systems are seen by many as "superior" to those operated by people because of the reduced risk of human error (Watts, 1991).

Human Operators and Inhuman Tasks

Organizations operating in an environment of high technical or operational risk ultimately depend for their performance upon human operators. The organizational design is geared to producing an individual who will always make the "right" decision in a critical situation (Rochlin, La Porte, and Roberts, 1987). Therefore, if an accident does occur, and is not beyond the operator's perceived range of control, the organization can reach one of two conclusions: (1) the organization failed in its design, placing the right operator in the wrong situation, the wrong operator in the given situation, or any operator in an impossible situation; or (2) the operator failed as an individual, despite or in violation of organizational design, training, and adequate rules. It is clearly more desirable for the organization to believe in (2), the fallibility of the individual, than in (1), its own imperfections.

The classic terminology for (2) is "pilot error," which in itself is a sufficiently broad term to warrant further explication. In the early days of passenger flying, it was assumed that there was a fair degree of risk involved, and that the causal factors were not only uncertain and difficult to anticipate, but perhaps unknowable in advance of actual circumstances. The job of the pilot was not just to provide technical skill as a machinery operator, or complex guidance by means of skills and techniques not generally known, but also to cope with the unanticipated, and threatening, as it occurred (Woods, 1987). Thus, when accidents did occur, the blame was almost always placed on "operator error"—except when the physical or technical malfunction was so blatant and so unambiguous that no action by the operator could possibly have saved the situation (Rasmussen, 1990).

In the case of large, complex organizations operating increasingly sophisticated technologies, the category of individual error as just described has to be further expanded to allow for the fact that some operations must be performed by a group of people acting in close concert. Thus, "pilot error" can be extended to "group error," where the group in question can be the team in the cockpit of an advanced airliner, the personnel on the bridge of a large ship, or the team in the Command Information Center on an *Aegis* cruiser (Rochlin, 1991).

Even when so expanded, "pilot error" remains a portmanteau, subsuming two general categories of presumptive malfeasance: failure to operate the equipment properly, or skillfully, or to follow various procedures and rules established to guarantee operational safety; and failure to rescue operations from an unanticipated or abnormal situation that was nevertheless within the presumed skill and capacity of the expert operator to rectify or remedy. In traditional terms, these might be divided into errors of commission and errors of omission. The two ways of parsing errors, however, fail to call attention to the central role of formal knowledge in organizations dealing with activities where anticipation, prediction, and modeling of equipment and behavior are central activities.

We can better deal with these two categories by following the terminology of Martin Landau, whose separation of errors is based on the epistemology of knowledge (Landau and Stout, 1979). In these terms, a "Type I" error is homologous to rejecting as false a hypothesis that is true—overlooking, ignoring, or misunderstanding the information presented even when it occurs within the envelope of the predicted or anticipated flow of events. It should be noted that this class of error also includes "anticipatable" equipment malfunction or stress over a wide range—such as engine failures, electronic malfunctions, or high-stress operational periods such as take offs or surface combat. Included in this class, therefore, is the failure to rectify or save a variety of situations that are presumed to lie within the range of an expert operator to correct or act upon.

The second is a "Type II" error, homologous to accepting as true a hypothesis that is false—accepting as true, accurate, or significant information that is misleading, incorrect, or irrelevant, or, by extension, projecting into a situation "external" beliefs or assumptions about the nature of the situation or the state of the system. Recent examples have included failure by pilots to set flaps correctly for takeoff or to note that de-iced wings have re-iced; believing that a gauge indicating that a nuclear power plant is losing water from a pressurizer is malfunctioning; interpreting radiation gauges at their maximum readings at face value rather than as pinned from overload; and persistent misidentification of civilian aircraft heading toward a naval unit in a combat zone as a military one preparing to attack.

However, even this more rigorous and formal scheme of classification is growing increasingly inadequate for large-scale technical systems, particularly those involving some degree of actual (physical) risk. The systematic progression and complexification of advanced technology have resulted in incidents that were once classified as Type I errors progressively moving toward Type II. Redundancies in equipment, presumptive higher mechanical reliability, and considerable sums spent on system design and large-scale integration result in the anticipation that such failures will be increasingly rare. Moreover, as will be discussed a bit later, a new class of error—"Type III"—is emerging, involving situations in which no reasonable hypothesis can be generated or evaluated at all in the available time with the available information.

In some cases, the very complexity of the equipment makes it quite difficult to ascertain what is going awry, why, and, at times, whether something is going badly or incorrectly at all. The responsiveness of modern, automated equipment may also allow a relatively minor error to escalate before the operators can intervene. As the equipment gets more complex, and presumably more reliable, the operator's job is also progressively redefined from that of error-detector and manager of continually suspect equipment to that of an "administrator" of integrated systems whose internal functions are remote from his or her experience (Roth, Bennett, and Woods, 1987). The supreme irony is that such a change in attitude and approach is often regarded as beneficial, even empowering (Zuboff, 1984).

Over the past twenty years, the repeated failures of complex operations involving sophisticated technologies have drawn considerable public and analytic attention. Based on an analysis similar to that in the preceding paragraphs, and on the formal literature on the sociology and structure of complex organizations, several analysts have even come to the uncomfortable conclusion that many organizations, including some that manage technologies or activities that might expose the public to serious harm, are now so structured that systemic failures of serious import are, if not inevitable, at least built into their very fabric. And one of the key elements in their analysis may be interpreted, in the terms of this chapter as a lack of essential friction.

Toward A Taxonomy of Error

Most of the literature on the failure of complex organizations managing complex, high-hazard technologies tends to focus on the most extreme consequences of organizational error. The literature provides an adequate taxonomy, but by focusing too closely on the nature of the organization and the special failures of its tasks, most of it fails to provide adequate grounds for systemic and structural analysis. What can be agreed upon in the majority of these cases is that the source of the failures was indeed internal to the organization itself; none were directly attributable to acts of violence, nature, or any other external agent.

Extant case studies can be broadly divided into:

a. Organizations that failed to adapt to changing technology, and retained traditional modes of decision-making (Reason, 1989; Shrivastava, 1987).
b. Organizations that adapted too rigidly, and shifted over completely from traditional to 'modern' (engineering-oriented) modes while casting off their experiential base.[6]
c. Organizations that failed to understand that changes in mission or organizational goals could undermine or render ineffective previous mechanisms for error control (NASA, especially in the case of *Challenger*: see, for examples Romzek (1987), Vaughan (1990)).
d. Organizations that neglected reliability or underestimated the impacts and/or consequences of error (Three-Mile Island; the *Exxon Valdez*; Reason (1989) and Turner (1978) cite many similar cases in their extensive work).
e. Organizations that failed to devise adequate rules and procedures, or to fully understand the importance of compliance (Chernobyl).

But given that a 'failure' occurred, or was perceived to have occurred, at what conceptual and systemic level did it take place? Whether these failures occurred at the individual level, at the group level, or at the administrative/ managerial level, they must have entailed actions (or erroneous inactions) by one or more individuals. In traditional analysis, a great deal of the current literature, and most newspaper or media reports of accidents or disasters, the central problematique is the determination of who caused the event. Rarely do such examinations extend to trying to assess whether there were underlying causes such as poor training, impaired judgment, or tension.

Too often, even these more general investigations continue to search for a human cause, providing a convenient modality for shifting the blame from the operators to other individuals upon whom they depended without acknowledging that a system can fail in the absence of culpability. Parsing for the conceptual and organizational bases of error is more difficult; in practice, it is rarely performed in the constant search to treat errors as lapses of responsibility, judgment, or even morality (e.g., drug use). Review boards constituted to seek blame will find some bases of error, even though many of the cases in the preceding taxonomy are examples of systemic rather than individual shortcomings.[7]

Some of the newer literature on failures and "disasters" takes this more systemic approach. Charles Perrow suggests that uncontainable errors become inevitable for organizations using technologies of a certain degree of complexity and potential hazard in situations where technological, regulatory, or organizational demands make for excessively "tight" coupling between action and feedback (Perrow, 1986). James Reason and colleagues argue further that some complex, tightly coupled organizations have modes and modalities of error designed into them from the outset (Reason, 1989; Rasmussen and Batstone, 1989; Woods, 1987). In such cases, the propensity for error may be inherent in the organizational design; described by Reason as "resident pathogens," they are not so much created by events as inadvertently released.

Karl Weick (1987) argues further still that we need to distinguish errors from mistakes: An "error" occurs when a person strays from a guide or a prescribed course of action, through inadvertence, and is blame-worthy, whereas a "mistake" occurs when there is a misconception, mis-identification, or misunderstanding. In this lexicon, then, errors are lapses in judgment or performance in a decision environment where time and information are adequate and ambiguity low. Mistakes arise from specific, individual judgments made in a context where the information was pre-sented poorly, or presented in such a way as to preclude the operator from exercising informed, expert judgment in the time, or with the information, available—what Weick has referred to as "errors of rendition," and what I categorize as falling within the new domain of Type III errors.

Friction and the Active Control of the Proclivity to Err

There are nevertheless many organizations that do perform tasks of technical complexity and potential hazard similar to those that have so dramatically failed in the past. If they are largely unnoticed, that is probably because a failure to fail is not quite so newsworthy as a dramatic failure. But there is a growing body of literature on the subject. My colleagues and I at Berkeley have for the past several years been studying three of these— air traffic control, naval flight operations, and utility grid management— in search of some of the common characteristics that distinguish these organizations from others (Rochlin, 1992).

Among our major findings is that the importance for reliability and safety of "social friction"—negotiation, consultation, and internal communication in operations—increases as the technology and system become more complex (La Porte, 1988; Roberts and Gargano, 1990; Rochlin, 1992; Rochlin, La Porte, and Roberts, 1987). Ironically, many of these adaptive strategies closely match the description of those aspects of industrial organization that Taylorism was specifically designed to combat, and that the neo-Taylorists are once again assaulting as wasteful and inefficient.

Functional adaptations include such classic increasers of organizational friction as the "inefficient" creation and nurture of informal decision networks and multiple informal organizational forms, even those designed to meet specific urgent tasks when and where they appear (Rochlin, La Porte, and Roberts, 1987). Another example is the explicit unwillingness of operators and operational managers to adopt or introduce new technologies or otherwise disturb the work environment unless absolutely necessary, for fear that they will disturb or disrupt the social formations that underlie successful operations.

In all three organizations under study, the operators have characterized their sense of having proper command and integration of information flows and system status as the equivalent of "having the bubble" in a ship's tactical command center (Roberts and Rousseau, 1989). When one has the bubble, all of the charts, the radar displays, the information from console operators, and the inputs from others and from the senior staff fall into place as parts of a large, coherent picture. However, given the large amount of information and critical nature of the task, keeping the bubble is a considerable strain. On many U.S. Navy ships, operations officer shifts are held to no more than two hours. "Losing the bubble" is a serious and ever-present threat; it has become incorporated into the general conversation of operators as representing a state of incomprehension or misunderstanding even in an ambiance of good information.

This notional bubble is one of the key elements in obtaining high and reliable systemic performance in organizational settings that involve considerable risk, large and basically uncontrollable uncertainties, and a high degree of organizational as well as technical complexity, operating in real-time situations where decisions with irrevocable consequences must be made quickly, in real time, on the spot (Rochlin, 1989). At the cognitive level, the conception of the bubble makes it possible for Navy operations officers, air traffic controllers, and (in a somewhat different context) experienced pilots to integrate a system of vast operational, technical, and cognitive complexity into a single, mental, representative spatiotemporal picture or map that organizes and orders the flow of information and allows decisions to be made on the basis of overall systemic situations and requirements.

The organizational costs are not negligible. As might be expected from the preceding description, bubbles cannot be acquired quickly or simply, and their establishment and maintenance require elaborate and overlapping networks of communications. Aboard the ships we have observed, tactical operations shifts overlap by up to an hour to make sure that the bubble is transferred smoothly to the succeeding tactical officer without a potentially dangerous break in routine or perception. In flight operations, too, there are many different observers, in different locations, following and analyzing the same set of activities, and communicating constantly about what they are seeing and how they are interpreting it. Similar procedures are followed in air traffic control and grid management. Even in more conventional and less pressing systems, such as nuclear power plant operations, the procedures for overlapping and communicating information during shift changes are extensive. Bubbles, then, are primarily human decision systems, maintained by constant human interactions with considerable redundancy.

Computers as Lubricants

On February 24, 1989, United Airlines flight 811 from Hawaii to New Zealand suffered major damage when a door blew off in a thunderstorm, taking with it a large chunk of fuselage (Fisher, 1989). Nine people were sucked out, twenty-seven were injured, and severe structural and control damage was sustained. Fortunately for those aboard the 747, Captain David Cronin was one of the older and more experienced pilots in the fleet. Relying primarily on his judgment and thirty-eight years of accumulated experience, he managed to retain control of the aircraft by "feel" and bring it safely back for a gentle landing, a feat that was regarded as near-miraculous by those who examined the airframe afterward. Will younger pilots, trained to fly safer and more reliable aircraft with automated control and navigation systems that provide little intellectual or tactile feedback, still be able to invoke their experience or "feel" some ten or twenty years hence? (Stix, 1991). Fortunately, this concern has spread. Even within the industry there is growing criticism—based at least partially on the grounds of experiential learning—of those who design control systems that require placing arbitrary limitations on what the pilot can make the aircraft do (Phillips, 1992).

Analysis of similar cases in other large-scale technical systems, particularly those with a high degree of automation, reveals the growing prevalence of operational complexity as a causal factor in accidents and failures. As the equipment becomes ever more elaborate, and the operators increasingly move from hands-on to representational modes of analysis, there is always the danger that the representation presented to the operator is in itself incorrect and/or misleading.

The shooting down of an Iranian airliner by the *USS Vincennes* in the Persian Gulf in 1988 is one such case, with tragic results (Rochlin, 1991). This ship was one of a new class equipped with very advanced and sophisticated information gathering and data-processing technology. Indeed, the entire mission of cruisers of this class and their *Aegis* target-acquisition and tracking system is predicated on the comparative rapidity with which the automatic equipment can recognize and distinguish radar images.

The proceedings of the board convened by the U.S. Navy to investigate the incident show little sign of any primary role for "bubble" formation aboard *Aegis* cruisers (U.S. Department of Defense, 1988). Instead, the role of the multiple technical systems, which heretofore served primarily to feed information *to* the tactical operations officer, seems to have been reversed. Aboard the *Vincennes*, the *Aegis*-directed, computer-operated anti-aircraft missile system is the primary system, with the ship's captain primarily involved in deciding whether to activate its firing mode or not. In this situation, the operator has become an input, not an integrator, and the captain an authorizer rather than a commander of action. In short, Fordism has come to the Navy, one of the last repositories of craft skills and the apprenticeship method in the United States.[8]

The historical trend toward more complex error sequences in increasingly complex systems may be described from two different but overlapping perspectives. At the level of responsibility, the traditional assignment of error to individuals has expanded to include, first, collective error at the team, group, or "system" level and, then, systemic error in which the essential cause lies with the way the framework and information for decisionmaking are structured or presented. At the level of interpretive meaning, Type I errors remain primarily the responsibility of the operator or the operating group, whereas the causes of increasingly common Type II errors increasingly lie at the level of system design and operational integration and administration. More notable is the growth of Type III or representational errors under conditions where the testing of existing hypotheses or the generation of new ones becomes difficult and uncertain. System failures may then occur even when the organization is making every attempt to guard against *both* of the two classic types of error (Reason, 1989). Indeed, some would argue that the effort to guard against errors in poorly designed complex systems can actually make them more dangerous because of the attendant increase in complexity and coupling (Perrow, 1984).

Traditional forms of error will of course persist. Operators may fail to follow rules, or to operate the equipment safely or correctly. The task may be such that exaggerated assumptions were made as to the operators' ability to take action to avoid an unanticipated situation. In many circumstances small-group errors may also be treated fairly accurately as extensions and generalizations of individual behavior, with an expanded understanding of the role of the group in forming and accepting false hypotheses and interpretations. But operators may well be placed in positions where there is no reliable post hoc way to evaluate whether they *could* have acted correctly within the parameters of a specific situation with the technology, equipment, and decision and control frameworks they were given.

The assertion that operators "could have" or "should have" been able to properly detect, interpret, and remedy an incipient error carries with it the embedded assumption that the organization has provided a correct representation, sufficient training and experience to judge whether or not the equipment is reliable, information appropriate to the circumstances, and a decisionmaking framework in which the operator can separate judgment of immediate risk from the well-being of the organization as a whole.

Systemic failure is increasingly being considered as alternative diagnostic. But too often the conclusion is that human beings are the weak link in systems performance, inherently incapable of operating without help. The "solution" then, is to introduce extensive automated system or process management and/or "expert systems" or other computerized or computer-based tools as modalities for offsetting or replacing operator judgment and the social processes of reliability instead of merely augmenting them.

The effort to reduce all errors instead of focusing on serious ones and tolerating lesser ones is problematic even in situations involving traditional errors of omission and commission. Even the most safety-critical system has a spectrum of possible errors and failures, ranging from the trivial to the systemically destructive. The occurrence and subsequent detailed causal analysis of error may be terribly frictional and highly consumptive of organizational time and resources, but organizations that do operate well deem them essential contributors to overall system safety. Few processes of human learning are as effective as trial-and-error, particularly when accompanied by extensive social mechanisms for analyzing and correcting error rather than trying first to assign blame. Moreover, automated and computer-based systems lack "common sense": They can draw only on the formal knowledge already encoded into their databases. When such systems are faced with a situation outside their range, failure to provide an answer at all may be the best outcome.[9] Only experiential knowledge can offer a reliable and sound basis for responding to unanticipated situations (Dreyfus and Dreyfus, 1986).

It is not necessarily the case that the drive for greater organizational efficiency is linked to the adoption of procedures designed to control error, even when they entail processes that increase organizational friction. Where this tension has been manifest, it tends to correlate with the introduction of modern computer technologies, including "artificial intelligence" technologies, by administrators and managers more familiar with the managerial technology than with the operations themselves. Nor is the primary reason necessarily to increase operational safety and reliability, although that is often invoked as part of the public argument. In almost every instance, from military operations to air traffic control, those introducing the new technologies argue that the operational *capacity* of the system is being severely limited by the friction of the required social processes of information tracking and checking and the slowness of social processes of redundancy.

In the search to eradicate errors and increase operational efficiency while reducing costs, it is very tempting for management scientists and middle managers trained in and familiar with computers and other advanced information processing and control technologies, to substitute them for operator discretion and judgement (Kling, 1980). As with every other historical aspect of scientific management, the outcome on paper is smoother operation, free of many traditional sources of friction. The price is the removal of the prime source of operator learning and experience, often accompanied by the decrease of morale that so often follows reduction in discretionary power and authority.

With these may also be lost not only the experiential knowledge that allows operators to anticipate crises but the sense of responsibility that motivates them to act on that intuition even at the risk of being criticized for the costs involved. Sliding smoothly and frictionlessly along in normal operation, a safety-critical organization so reconstructed may well be gliding towards the edge of an abyss, with no social sensor system to detect the gradual change in terrain and no one empowered or experienced enough to slow it down in time to prevent a major catastrophe.

Negentropy and the Creation of Disaster

In his studies of 'man-made disasters,' Turner (1976, 1978) expanded the categories of organizational failure beyond the merely responsive. Organizations are by nature 'negentropic'—designed to order, structure, and manipulate their environment (Katz and Kahn, 1966). Organizational outputs may therefore be characterized as the product of energy and information. Turner points out that the information can be wrong and the product destructive: "disaster equals energy plus misinformation." The very processes of assembling energy and information to achieve a desired goal can easily result in a very good organization organizing to produce a very bad effect. In general, Turner argues, organizations are more susceptible to such effects when a large, complex problem is dealt with piecemeal by different units or subunits (even when formally related), very much along the line of Reason's description of the origins of "resident pathogens."

Large-scale system failures entailing considerable harm—disasters—may grow from seemingly ordinary and common errors, particularly when there is nothing to damp or interfere with the propagating causal chains. The chain of events preceding the initial phases of the accident at Three-Mile Island (Sills, Wolf, and Shelansky, 1982), the preparations for turbine testing at Chernobyl (Medvedev, 1991), and the O-ring controversy prior to the launch of the *Challenger* (Romzek and Dubnick, 1987; Vaughan, 1990) shared similar properties. In each case, the "frictional" processes of organizational oversight and control failed. But that fact alone might not have served to overcome the more conventional technical and operator error-control mechanisms.

These three cases also shared to varying degrees the negentropic tendencies identified by Turner. Organizational and managerial mechanisms designed to ensure order, to manage and reduce uncertainty, and to increase efficiency helped magnify the results of untimely, incomplete, or incorrect information into major disasters, creating erroneous "anti-tasks" that the organizations then executed with its usual diligence. Similarly, the mis-identification of the Iranian airliner by those on the USS *Vincennes* became such a central tenet and organizing principle of subsequent behavior that it could not be successfully challenged even when controverting evidence became available through the normal process of error checking (Rochlin, 1991; U.S. Department of Defense, 1988).

Organizational friction as a process by which decisions and actions are subject to continuous and constant discussion and review, preferably by units independent enough to form their own judgments about the course of events, becomes more critical when the possibility of negentropic processes is high. The Berkeley research group on organizational reliability has observed such mechanisms at work in all three organizations under study. Remarkably, process implementation is "self-conscious" in all three cases, even though those responsible for it lack the formal language for justifying the apparent waste of organizational resources.

In the case of naval flight operations, there is a constant and quite redundant process of checking and cross-checking that has, if anything, been strengthened in recent years as part of a conscious attempt to reduce the frequency of consequential errors (Rochlin, La Porte, and Roberts, 1987). In the one nuclear power plant we studied in depth, there is an active cultivation of two different perspectives for diagnosing problems to provide for independent analytic paths if something is not behaving properly (Suchard and Rochlin, 1992). While in air traffic control, the "paper trail" of slips that carry aircraft identification and information is still maintained in parallel with the electronic consoles (La Porte, 1988).

In each of these cases, however, there is a manifest and imminent threat to the organization's present way of operating that arises from the diagnostic opinion that the duplication is not only wasteful of human and fiscal resources but actually slows down operations. In the search for increasing operational efficiency, measured as output of the desired good (whether it be kilowatt-hours of electricity, number of sorties, or density of air traffic), there is for each organization some external pressure to introduce computer-based decision mechanisms or automated modes of operation.

As with most automated equipment, there is no fundamental reason for assuming that the introduction of decision aids and new methods and modalities of managing information and providing data will necessarily decrease operator involvement or the role of operator judgment. Indeed, the entire rhetoric surrounding the use of "expert systems" has been much moderated in recent years, tending away from absurd claims of replacing human judgment in favor of assisting and informing it (Winograd and Flores, 1986). But so long as the main purpose of such equipment is seen as increasing operational tempo, its introduction will almost surely lubricate the social processes of self-organization, while decreasing the time available for internal review and oversight.

Given the inherent limitations of human recognition and reaction time and the rapidity and data-processing ability of even a collection of human actors, there is some doubt as to whether the human-mediated damping processes I have characterized here as "essential friction" can be applied to automated systems in sufficient degree or in sufficient time to have much effect on outcomes. If, as I argued earlier, the equipment is deployed in the traditional neo-Taylorist fashion that seeks to prevent small errors and convey experiential knowledge to the operators, it is also likely that the causes and consequences of propagating error chains will not be observed or interpreted until the organization has magnified them to major proportions.

Conclusion: The Essentiality of Friction in A World of Risk

The work of Reason, Turner, Weick, and others studying the failures of complex technological systems tends to support Perrow's assertion that some of the major technological disasters of recent years are indeed examples of "normal" accidents: normal in that the complexity and coupling of organization and task make interpretation difficult and outcome uncertain and perhaps unpredictable; normal in that the need for quick and accurate diagnosis, interpretation, or intervention exceeds operator capabilities; normal in that the mechanisms by which errors or mistakes multiply into accidents may be built into the organizational or system design. Although the Berkeley group may seem to go against the grain in arguing that there do exist some organizations just as complex and tightly coupled that manage to arrange themselves to counter such tendencies — often self-consciously — it is difficult in the present environment to be sanguine about their ability to maintain those carefully created and tended error-correction and detection mechanisms.

Returning to the central metaphor of this book, I should add that my critique of many of the modern breed of scientific managers and consultants has to do with their continued adherence to an outmoded managerial and social ideology based on overly mechanical nineteenth-century beliefs

(Rabinbach, 1992). In physical terms, the application of a force over time (energy) should result in motion; through disorder, friction converts some of that energy to heat (entropy), reducing the motion or bringing it to a halt. If the output is judged only in terms of motion, friction should be decreased whenever and wherever possible. To those whose worldviews are mechanistic (and even scientistic), entropy equals disorder, and friction equals waste, and both disorder and friction are social and ideological negatives to be sought out and eradicated.

Such worldviews persist among many "scientific" or "econometric" managers, particularly those specializing in the more technical aspects of operation. An organization or firm is given a series of resources (time, money, people) as inputs and is expected to produce for it one or more productive or regulatory outputs. The social processes that confer experience, reliability, or safety consume some of these resources while producing no measurable output. Therefore, they are "frictional" (i.e., wasteful) and to be identified, isolated, and eliminated as part of the process of system "optimization." Worse yet, they are disorderly. Even when the relative costs are low, they remain suspect because their outputs are uncertain and they cannot be readily subjected to central oversight and control.

The mechanistic analysis that characterized the excesses of traditional Taylorism and Fordism focused primarily on the number of widgets being produced rather than on their quality or the reliability of the processes and procedures of production. If quality was low, errors were high, or the line shut down frequently due to the neglect or deliberate actions of the laborers, those were regarded as external factors to be combated and defeated separately, by means varying from social coercion to direct threats or termination. But there were and are limits to the effectiveness of these tactics, and the problems I've described persist in more traditional manufacturing firms even now—the U.S. automobile industry being a notable source of examples.

One of the primary accomplishments of post-Taylorist management theory was the recognition that productivity is multidimensional, and that social factors and social constructions are central elements in maintaining an effective organization with high-quality outputs (Merkle, 1980; Waring, 1991). Moreover, it is now common to recognize that a proper measure of productivity must also encompasses quality; tradeoffs between quantitative and qualitative aspects of production and performance are necessary and not just contingent if firms are to remain competitive in today's world markets.

The stakes are considerably higher where safety and risk are involved. The number of large, complex organizations responsible for managing, regulating, and controlling the hazards associated with modern industrial and technological systems continues to grow steadily. Whether operating oil rigs in the North Sea or overseeing chemical plants in heavily populated areas, controlling air traffic or operating aircraft, these organizations can never be judged successful unless they continue to promote and maintain public safety and well-being. Yet they, too, are subjected to constant pressure to increase productivity and efficiency—even while maintaining, or even increasing, safety and reliability.

Properly defined and implemented, computers and other automated systems, even when rated as "intelligent" or "expert," are nothing more than tools. Their real power lies in the unprecedented flexibility and adaptability they bring to the definition and execution of task and process. There is no reason in principle why they cannot be used to increase the essential frictional processes of safety-critical as well as ordinary organizations; to search for and help identify and damp out the consequences of errors; to provide the interpretive and contextual richness that minimizes opportunities for misperceptions, misapprehensions, and other sources of organizational mistakes; or to slow down the processes of decisionmaking whenever there is some ambiguity about the range of possible outcomes (Carroll and Reitman, 1987; Winograd and Flores, 1986).

Given the present search to eradicate human error, it is more likely that computers will be used primarily as mechanisms for minimizing human involvement in the decision loop—eliminating small errors for the sake of productivity, speeding up decision processes in the name of efficiency, and making integrated, systemic decisions on the basis of complex models whose parameters and dimensions are not clearly understood even by those who constructed them. In the traditional search to reduce and eliminate "social friction" and "wasted" time, such procedures suboptimize, bringing about short-term productivity and reliability at the expense of increasing the risk of serious failures over the long term.

At stake here is not so much the appropriateness of the metaphor but our enslavement by it, the persistence of the historical context in which the negative affect of terms such as *friction* and *entropy* has been socially constructed. The mechanistic worldview of the nineteenth century that transformed efficiency from a technical term to an ideology, carried into the managerial realm by the excesses of Taylorism and elevated to a pseudoscience by the blinder disciples of operations research, systems analysis, and systems engineering, is remarkably persistent among those responsible for developing, deploying, and implementing the new methods and modalities of computer technologies.

Since the dawn of the computer revolution, Western society has been haunted by the fear that we may have created something beyond our control or, possibly, our understanding. From HAL, the mad computer of Kubrick's *2001*, to the madness of *Westworld* or the inhuman and impersonal persistence of the liquid metal robot of *Terminator 3*, such images remain largely in the traditional mode; the dominant fear is loss of autonomy, loss of authority, loss of the power to initiate or direct events.

What I address in this chapter is another concern less obvious, less salient, and more relevant: the possibility that what is being lost in the implementation of the new intelligent technologies is not the authority to initiate events but the ability to identify, evaluate, and control the consequential chains that ensue. The prospect is not so much that of a society being dominated and ordered by the automata as that of a society slipping and sliding from one self-organized and self-created crisis to another, lacking the social and organizational mechanisms to dissipate accumulated momentum or the time to evaluate and counter adverse negentropic processes.

For most of this century, what has been at stake in the search for organizational efficiency and administrative control is command—who has the power to say "yes," and to what, who has the power and authority to rule. But what must be recognized in the modern era is that it is at least as important to consider who retains the power to negate, to dissipate, and to terminate. Friction is essential in organizations because it provides space for the exercise of the ultimate human responsibility—the power to say "no," and to enforce it.

References

American Academy of Arts and Sciences. "Risk." *Daedalus*, 119, No. 4 (1990).

Anonymous. "Our Do-Nothing Government." *The New York Times*, March 30, 1992, A15.

Barnard, Chester. *The Functions of the Executive*. Cambridge, Mass.: Harvard University Press, 1938.

Beniger, James R. *The Control Revolution: Technological and Economic Origins of the Information Society*. Cambridge, Mass.: Harvard University Press, 1986.

Blau, Peter M., McHugh Falbe, Cecilia, McKinley, William, and Tracy, Phelps K. "Technology and Organization in Manufacturing." *Administrative Science Quarterly*, 21 (March 1976), 21-40.

Brissy, Jaques F. "Computers in Organizations: The (White) Magic of the Black Box" (pp. 226-236). In *Organizational Symbolism*. Ed. Barry A. Turner. Berlin: de Gruyter, 1989.

Carroll, John M., and Reitman, Judith. *Mental Models in Human-Computer Interaction:*

Research Issues About What the User of Software Knows. Washington, D.C.: National Academy Press, 1987.

Chandler, Alfred D., Jr. *The Visible Hand: The Managerial Revolution in American Business*. Cambridge, Mass.: The Belknap Press, 1977.

Cyert, Richard M., and March, James G. *A Behavioral Theory of the Firm*. Englewood Cliffs, N.J.: Prentice-Hall, 1963.

Danziger, James N. *People and Computers: The Impacts of Computing on End Users in Organizations*. New York: Columbia University Press, 1986.

Doray, Bernard. *From Taylorism to Fordism: A Rational Madness*. Trans. David Macey. London: Free Association Books, 1988.

Douglas, Mary T., and Wildavsky, Aaron B. *Risk and Culture: An Essay on the Selection of Technical and Environmental Dangers*. Berkeley/Los Angeles: University of California Press, 1983.

Dreyfus, Hubert L., and Dreyfus, Stuart E. *Mind over Machine: The Power of Human Intuition and Expertise in the Era of the Computer*. New York: The Free Press, 1986.

Fisher, Lawrence M. "Experience, Not Rules, Led Airliner Crew in Emergency." *New York Times*, March 4 1989, p. 1.

Hirschhorn, Larry. *Beyond Mechanization*. Cambridge, Mass.: MIT Press, 1984.

Howard, Robert. *Brave New Workplace*. New York: Viking, 1985.

Hughes, Thomas P. *Networks of Power: Electrification in Western Society 1880-1930*. Baltimore: Johns Hopkins University Press, 1983.

Kates, Robert W., Hohenemser, Christoph, and Kasperson, Jeanne X., Eds. *Perilous Progress: Managing the Hazards of Technology*. Boulder, Colo.: Westview Press, 1985.

Katz, Daniel, and Kahn, Robert L. *The Social Psychology of Organizations*. New York: John Wiley & Sons, 1966.

Kittel, Charles. "The Nature and Development of Operations Research." *Science*, 105 (January 1947), 150-153.

Kling, Rob. "Social Analysis of Computing: Theoretical Perspectives on Recent Empirical Research." *Computing Surveys*, 12, No. 1 (March 1980), 61-110.

La Porte, Todd R. "On the Design and Management of Nearly Error-Free Organizational Control Systems" (pp. 185-200). In *The Accident and Three-Mile Island: The Human Dimensions*. Ed. D. Sills, C. Wolf, and V. Shelanski. Boulder, Colo.: Westview Press, 1982.

———. "The United States Air Traffic System: Increasing Reliability in the Midst of Rapid Growth" (pp. 215-244). In *The Development of Large Technical Systems*. Ed. R. Mayntz and T. P. Hughes. New York: Martinus Nijhoff, 1988.

Landau, Martin, and Stout, Russell, Jr. "To Manage Is Not to Control: Or the Folly of Type II Errors." *Public Administration Review*, 39 (March/April 1979), 148-156.

Medvedev, Grigory. *The Truth About Chernobyl*. Trans. Evelyn Rossiter. New York: Basic Books, 1991.

Merkle, Judith A. *Management and Ideology: The Legacy of the International Scientific Management Movement*. Berkeley/Los Angeles: University of California Press, 1980.

Nevins, Allan. *Ford: The Times, the Man, the Company*. New York: Charles Scribner's Sons, 1954.

Noble, David F. *America by Design: Science, Technology and the Rise of Corporate Capitalism*. New York: Oxford University Press, 1977.

———. *Forces of Production: A Social History of Industrial Automation*. New York: Oxford University Press, 1986.

Perrow, Charles. *Normal Accidents: Living with High-Risk Technologies*. New York: Basic Books, 1984.

Perrow, Charles. *Complex Organizations: A Critical Essay*. Third ed. New York: Random House, 1986.

Phillips, Edward H. "Pilots, Human Factors Specialists Urge Better Man-Machine Cockpit Interface." *Aviation Week & Space Technology*, 136, No. 12 (March 23 1992), 67–68.

Rabinbach, Anson *The Human Motor: Energy, Fatigue and the Origins of Modernity*. Berkeley and Los Angeles: University of California Press, 1992.

Ralph, R., ed. *Probabilistic Risk Assessment in the Nuclear Power Industry: Fundamentals and Applications*. New York: Pergamon Press, 1988.

Rasmussen, Jens "Human Error and the Problem of Causality in the Analysis of Accidents." *Phil. Trans. R. Soc. Lond.*, B 327 (1990), 449–462.

Rasmussen, Jens & Batstone, Roger eds. *Safety Control and Risk Management*. New York: Wiley, 1989.

Reason, James. *Human Error: Causes and Consequences*. New York: Cambridge University Press, 1989.

Roberts, Karlene H., and Gargano, Gina. "Managing a High-Reliability Organization: A Case for Interdependence." (pp. 146–159). In *Managing Complexity in High-Technology Organizations*. Ed. Mary Ann von Glinow and Susan Albers Mohrman. New York: Oxford University Press, 1990.

Roberts, Karlene H., and Rousseau, Denise. "Research in Nearly Failure-Free, High-Reliability Systems: 'Having the Bubble.'" *IEEE Transactions*, 36 (1989), 132–139.

Rochlin, Gene I. "Informal Organizational Networking as a Crisis Avoidance Strategy: U.S. Naval Flight Operations as a Case Study." *Industrial Crisis Quarterly*, 3 (1989), 159–176.

———. "Iran Air Flight 655: Complex, Large-Scale Military Systems and the Failure of Control" (pp. 95–121). In *Responding to Large Technical Systems: Control or Anticipation*. Ed. Renate Mayntz and Todd R. La Porte. Amsterdam: Kluwer, 1991.

———. "Defining High-Reliability Organizations In Practice: A Taxonomic Prologomena." In *New Challenges to Understanding Organizations*. Ed. Karlene H. Roberts. Beverly Hills: Sage, 1992.

Rochlin, Gene I., La Porte, Todd R., and Roberts, Karlene H. "The Self-Designing High-Reliability Organization: Aircraft Carrier Flight Operations at Sea." *Naval War College Review*, 40, No. 4 (Autumn 1987), 76–90.

Romzek, Barbara S., and Dubnick, Melvin J. "Accountability in the Public Sector: Lessons from the Challenger Tragedy." *Public Administration Review*, 40, No. 3 (May/June 1987), 227–238.

Roth, E. M., Bennett, K. B., and Woods, D. D. "Human Interaction with an 'Intelligent' Machine." *International Journal of Man-Machine Studies*, 27 (1987), 479–525.

Scott, W. Richard. *Organizations: Rational, Natural, and Open Systems*. Englewood Cliffs, NJ: Prentice-Hall, 1981.

Shrivastava, Paul *Bhopal: Anatomy of a Crisis*. Cambridge, MA: Ballinger, 1987.

Sills, David, Wolf, Charles & Shelansky, Victor, eds. *The Accident and Three-Mile Island.* Boulder, CO.: Westview Press, 1982.

Simon, Herbert *Administrative Behavior: A Study of Decision-Making Processes in Administrative Organizations.* 3d ed. New York: The Free Press, 1976.

Stix, Gary "Along for the Ride?" *Scientific American* (July 1991), 94–106.

Suchard, Alexandra & Rochlin, Gene I. *Operators and Engineers: Two Perspectives on Nuclear Power Plant Operations.* 1992.

Taylor, Frederick Winslow. "The Principles of Scientific Management." In *Classics of Organization Theory.* Ed. Jay M. Shafritz and J. Steven Ott. Chicago: The Dorsey Press, 1987. pp. 66–80.

Thompson, James D. *Organizations In Action.* New York: McGraw-Hill, 1967.

United States Department of Defense *Investigation Report: Formal Investigation into the Circumstances Surrounding the Downing of Iran Air Flight 655 on 3 July 1988.* Washington, D.C.: United Stated Department of Defense, 1988.

Vaughan, Diane. "Autonomy, Interdependence, and Social Control: NASA and the Space Shuttle *Challenger.*" *Administrative Science Quarterly,* 35 (1990), 225-257.

Waring, Stephen P. *Taylorism Transformed: Scientific Management Theory since 1945.* Chapel Hill, N.C.: The University of North Carolina Press, 1991.

Watts, Susan. "Computer Watch on Nuclear Plant Raises Safety Fears." *London Independent,* October 13, 1991, 1.

Weick, Karl E. *The Social Psychology of Organizing.* New York: Random House, 1979.

———. "Organizational Culture as a Source of High Reliability." *California Management Review,* 29, No. 2 (Winter 1987), 112–127.

Weinberg, Nathan. *Computers in the Information Society.* Boulder, Colo.: Westview Press, 1990.

Winograd, Terry, and Flores, Fernando. *Understanding Computers and Cognition.* Reading, Mass.: Addison-Wesley, 1986.

von Glinow, Mary Ann & Albers Mohrman, Susan *Managing Complexity in High Technology Organizations.* New York: Oxford University Press, 1990.

Woods, David D. "Commentary: Cognitive Engineering in Complex and Dynamic Worlds." *International Journal of Man-Machine Studies,* 27 (1987), 571–585.

Zuboff, Shoshana. *In the Age of the Smart Machine.* New York: Basic Books, 1984.

PART FIVE

Elation and Frustration

8

Playing, Writing, Wrestling

Sigrid Combüchen

Playing

The kitten is busy with her toy mouse. All pricked ears and wispy tail she gives her vivid performance of hunter and prey. She knows of course that she is handling a dead object. But in a corner of her mind lurks handed-down cat knowledge, which maintains that the very nature of play is to breathe life into things passive; by force of imagination as well as literally, with swift paws.

Her evening ritual takes place under the living room table, on the exact spot where the table legs are joined by a double cross-bar close to the floor. The kitten has chosen this spot for the sake of challenge. She has learned all there is to know about chasing balls and mice across polished floors. Minor obstacles posed by rugs or stairs hold no further excitement. To refine her skills she needs something double-fenced, triple-fenced, or multi-fenced. The wooden cross with all its nooks and corners and hard-to-get-at squeezy parts, gives "mice" ample opportunity to hide almost out of reach and pursuers have to somersault, bang their heads, kick, and get stuck.

Most intriguing from an observer's point of view is to see how the kitten gradually increases the level of complexity. The varieties of the chase grow more sophisticated, she outsmarts herself with new tricks and once she has got the knack of those, she immediately abandons them and proceeds on to something even wilder. Final success is in itself no goal, the real issue of the play is to maintain a balance, where her skills remain somewhat inferior to her task. It is her creative energies that are being trained, not her ego.

These games are of course preparations for future kills, but, as we know, her methods will prevail when she catches live creatures. A cat relaxing her hold on a prey—allowing it some hope just before the finale—brings to mind the question of human free will and moral choice within the Will of Deity. She does so like to encourage the primitive optimism of the mouse struggling for life, just to quench it whenever she wants to. Still, once she has killed and the challenge has gone limp, her sense of pride and superiority is instantly invaded by loss of concentration and utter boredom.

One year later she sits on the carpet, watching her litter repeat the performance. Her eyes are intently, coolly fixed upon them. The situation seems educative, but she takes no action. At the same time, but in other places, human parents, child-care personnel, siblings instruct children about the right way to play and live. Piling up arguments, adjusting hands and legs and postures. Occasionally taking over altogether for the fun and final ease of child-playing with grown-up skills. You see vacancy in the eyes of children brushed aside to be audience at their own games, where they were actually busy having their constructive failures, repairing them and raising the level of endeavor in order to fail once again.

These descriptions of idyllic circumstances have been made in order to idealize the existence of friction. Two key concepts hover within the picture above. The first one is challenge, a highly competitive word, suggesting success if you are cut out for it. Challenge is something you rise to; you go up on your hind legs and "jump." The hurdle may give you trouble and take its time, but once you're over it you have finished a certain task and put it behind you.

The other concept is inertia. In contrast to the clear peak moment of challenge, inertia has a soft, clinging quality. It goes with your system. Your physical and psychological being contains built-in inhibitions, as unforeseeable as the color of your eyes or the texture of your skin. Inertia is also outer phenomena seen from within—the dead weight and opaqueness of "things"—and it is interplay between inner and outer rigidity in everchanging combinations. In that way it is one of the constancies of life. It will seldom define itself by dramatic features. Metaphorically, it might compare to temperature, always present, sometimes changing, sometimes dulled into dusting heat or permanent fog. You must adjust to it, and most of the time you do so unconsciously, but occasionally there is a conflict between yourself and this law of nature. You will not in any definite sense win it, but you may gain or learn something from the friction between your cut-out personality and the impersonal forces that surround it.

When I sit down to ponder upon the concept of friction, several images introduce themselves. First of all, the green neutron star, matter collapsed into extreme density, rotating at great speed, thereby creating friction that sheds "showers" of energy. Any astronomer would shudder at so inaccurate and romantic a description. But a writer can use inaccuracies for metaphors, and none could shine brighter than a green star, pouring out radiance by way of its violent friction with—almost—vacuum.

Retreating from space to earth, and to the creations performed by humans, these could not take place in a vacuum or empty surroundings. We are not dense enough; we lack the speed. People are born out of slow, existing matter and have to create in relation to it. Your cheek will glow if you rub it or if something stirs your inside. Similarly, your body and soul will "radiate" only if it is touched or stroked or smitten by some influence. In the modern world, however, you push your way into surroundings that continually grow in density as urbanization increases and so-called information clogs the paths of human intercourse. These amounts of stimulants can on the one hand revive the human soul; but on the other congest into a nightmarish babble, where everything inventive or creative comes to a standstill. Not too dense, not too empty—this must be the ideal but indefinable measure of inertia versus man.

Sparks from cast iron against iron, when industry was young, feature the optimism of friction. Love's *labor* comes to mind as a singular source for radiation. But the most striking exposure of friction as a generative force is—connecting back to the kitten—sports and games.

The postures of sports refine into a literal ballet of friction. This may seem a contradiction in terms, since ballet itself defies the concept or the visibility of it. If you sit close you may see the sweat and hear the thuds and panting, but at a distance there is an ethereal quality, general for bodies trained to stay tense and keep their weight-point at a high level. In sports, although it generates aesthetic postures and movements, the body is not so much an instrument to penetrate the significance of those as a tool to meet opposition. The counterforce may be the other team, the clock, the track, or the equipment; the latter increasingly demands expert skill as it gives more sophisticated results. All of these are examples of *designed* friction. They initiate the wish to advance and then hinder ready advancement, keeping a benign conflict alive. Of course, teams and individuals do not have this philosophical viewpoint during contest. They definitely wish to end it, favorably. But Sport in itself wants equilibrium between equal parties, who develop the idea of the game somewhat beyond their known ability. And outside the narrow aspect of winners and losers, Sport always gets what it wants. From chess to water polo—offhand it is hard to imagine any other field where friction is so visibly built into human activity as a stimulating impediment.

The postures I referred to earlier can more readily be studies on stills than on motion pictures. In the flow of events on a football field, the moments suck each other up. They summarize into definition only when the referee blows his whistle for foul or a goal. Or when the photographer pokes his objective into the bustle and freezes a slice of it. What he will then generally choose is a group of two or three players, who simultaneously defend and advance, hampered by each other and by the additional friction of rules. In the picture one will see young men (mostly) with weary old faces and wiry old bodies, and their postures will have the beauty of forgetfulness—which is the exact opposite of the impudent poses of the sleek, narcissistic body.

The strategies of sports and games are often considered to be derived from military theories. This thought does make sense (especially if your attitude to sports is hostile).

When it comes to means you find similarities, at least metaphorically: the drawing up of tactics, the strangely cumulative effects of bad or good luck and how this "luck" thing affects what I here call friction, making it almost unpenetrable, alternatively almost doing away with it. Also there are in the worlds of both sports and soldiering such things as systematic training and "match training." In the first instant your own lack of skill and experience constitutes friction, and you "work" it through pounding repetition. Your environment is an enclosure, in contrast to match training, where general unpredictability creates friction (apart from the enemy action): the weather, the field, a body that inexplicably goes limp and lazy when alertness is what is most required from it.

Of course, a world of circumstances tells sports and war games apart. Despite the facts that athletes need to wear protective upholstering and supporter gangs take any opportunity to ruin cities, one must emphasize that sport is symbolic action—acted out bodily—and that, far from the ambition of permanence in military victory, the athlete is quite aware that success is as short-lived as a single ray of sun. He is a conqueror only in the tabloids. In real life he acts a principle of existence, where failure and the monotony of inertia are far more frequent than success and peak moments. The final image of the athlete, accordingly, is resignation. Flesh and bones can undergo hard cultivation for only a period of life—or even a period of youth. At a comparatively early age the athlete will give the deterioration of body and mind an observance that other people can postpone for decades to come. Every human being is a clearing in the forest that will eventually be overgrown, but most of us can ignore the minor weeds and accept some shrubbery. The elite must constantly tidy its clearing and still be overgrown in the end. That leads into locked contact with friction and the melancholy of early aging.

Writing

First, a general observation about friction in fiction: Any creative process will be initiated by a vision that seems amazingly clear. But as soon as it leaves its brooding-nest in your brain and starts a journey toward realization, clearness will blur into haze. Not only will you lose the outlines of your image but, worse, its general direction and the intense feeling of purpose that was originally there. What is the reason for this? Is there a reason for this?

The most natural conclusion would be that any wothwhile act of creation deals not only with its particular topics but with all major and minor quests that interplay at your present level of perception. In a sophisticated society, actual invention is rare. As an artist you can't create novelty—as postmodernism has taught. You could, however, "invent the wheel" and do it quite innocently and even manage to convey its image anew and in a striking manner. In this process the irony of repetition will still creep up, however flippant and trivial irony may seem to you. It brings on the message that today's invention is to "rephrase" what is already known, subtly enough to give a fresh and firsthand vision of it. (This rule of "firsthandness" and "freshness" mockingly applies even to postmodernists.) Here, the mentioned level of perception can create the worst inertia: Sense and nonsense from the media constitute a growing portion of what you currently know. If you give in to it you lose your bearings, if you disregard it, your work will seem slightly sterile. What you have to do is treat the flow of contemporariness with an organizing mind and give your full attention to its grinding against your own taste. Writing the right way seems to have very narrow possibilities, a block of frustration.

But of course there is such a thing as easy success. I will not claim that it is immoral in itself to slide without effort into the right place at the right time. All artists occasionally experience coincidence as God's Finger, pushing everything into position and arranging it in a manner that they could never have achieved through just work and willpower. A few geniuses appear to have tamed that Finger for their own benefit.

However, there is the slope of popular acclaim: the pace at which a restless culture creates and kills mock geniuses at a heretofore unknown rate. In these circumstances *appeal* plays a more important role than effort. The genius of Presence is a vital part of the modern success story, and in that context what you *are* is more important than what you *do*. There has always been less friction on the road for the voluptuous blonde with dazzling eyes than for the unpresuming mouse wearing sensible shoes and pink cotton underwear. This Life's Unfairness is one of the mysteries of life and has to be accepted as such. But when it turns into a system within public life and becomes more and more institutionalized, appeal turns into a patent shortcut to recognition and identity, in a way that achievement can hardly obtain any more.

Does appeal somehow eliminate friction? Obviously the smoothing of certain paths will shift friction from a specific individual to a sociological scale, in the same way that time pressure in the lives of sought-after personalities transfers the entire bulk of dead and senseless time onto discarded and unwanted people. Giving the matter a solely political dimension is a sham, though, suggesting administrative solutions to the problems of existence. In real life some people are born easy-goers and live fluently, unaffected by, or disregarding, obstacles. Others stay frictional and difficult, whatever efforts are made to adjust the world to them. The same kind of individual "difficulty" you can find in abstractions, in organizations and ideas. But as you do not wish to rid yourself of valuable problems, you look for ways to adjust to or, even better, make use of the friction they cause.

Within the creative process the first image is an equivalent of the hypothesis, and it equally has to be proven. Proving in this case, however, is not the same as finding evidence and harmonious facts. Instead, it could be described as forcing your vision through the vast somnambulence of your body and mind and creating flickers of alertness, which will eventually combine into something that you decide bears sufficient resemblance to your first inspiration.

One of the prerequisites of survival is a certain restfulness within the organism, something you could almost call ignorance of self. Charlie Parker said that if your mind actually registered everything you hear, you would go mad. So you build defenses against impressions, blunt your senses, and—in a way—protect your inertia. Of course there is notion, there is motion. There is reaction, observation, and even initiative going on all the time. But these activities keep mainly to their tracks, while the restful bulk tags along. People will or will not accept that large and vital parts of them are vegetative. But they will certainly experience the nature of their inertia whenever they attempt to transport a vision into practical existence.

From fancy to fact you face yourself as an adversary, and not necessarily in equal proportions. Your innocence of purpose may be opposed by your laziness, your sense of routine by loss of confidence, your ambition by sheer inability. You surprise yourself with the cool ingeniousness of your own "opposition": The road to fulfillment is winding, energy and concentration are being shed to a great extent, and the final picture has this very vague, compromised likeness with what you first intended. But the failure in bringing about your vision always has to be considered along with your success in bringing something to life at all.

The novel—one might say *each* novel—spans between poetic expression (corresponding with your basic idea) and the dreary triviality of transport (related to the somnambulence mentioned earlier). To juxtapose all the attitudes that find room between these limits gives the author ample opportunity to deal with "friction matter." These dealings could be sampled in many ways. I will focus upon two writers, who may or may not have contemplated the word *friction* but illuminate it from two angles—Doris Lessing, working through the nitty-gritty of existence, and Rebecca West literally (or literarily) penetrating the concept as such.

Politically, modern literature has two frontiers for the individual writer to waver between. One of them is—roughly—the direct approach to subject matter, whereby style is a fusion, a potent concentration subjected to the importance of what has to be told. Of the other one could say that it diffuses its theme by means of style. Sometimes this means that the artfulness of the prose remains an end in itself. Generally speaking, however, we have here an attempt to collect the same truth as in so-called realism. The difference is that one doesn't choose to take a photograph of the body in question but, rather, x-rays it or—to idealize a bit—even travels along its veins or nervous system with a micro-camera. This process may give an artful impression or even seem like literary domestication if the writer has a penchant for the beautiful. The conflict between the assumed down-to-earth attitude and Art is mostly verbal, since most novelists harbor a bit of each. Still, you have periods when *decision* excludes one or the other, whether this happens collectively—as in the 1970s—or on an individual scale.

Doris Lessing was never one to "write the world beautiful." An overall description of her work would define her as a philosopher, or as someone who extracts vital essence from political and sociological phenomena, then projects them clearly into our own time or slightly ahead of it. She is earnest in an unsolemn way and—without fuss or circumstance—examines the life and vanishing of people and ideas (and cats), allowing each character its individuality, without stressing it. Her style has shifted over the years, owing to her themes. Though seldom flamboyantly "literary," it has been adequate, and there have been metaphors and feelings of mystique and characterization and whatever else you need in a catching novel.

Her reason for writing *Jane Somer's Diaries*—the two novels sent pseudonymously to her regular publisher, who rejected them for lack of quality—seemed dubious at first. The cause célèbre was of course that the rejection was due to an unknown name and would not have taken place without the alias, which goes to prove etc. But Doris Lessing's "proof" was in fact more experimental than the media upheaval led one to believe. The *Diaries* are indeed *bad* literature according to all established rules of creative writing (though quite readable in an entertaining sort of way). Triviality is pursued trivially. A London career woman in her middle age gets entangled in the misery of helpless youth and aging, when she finds herself the victim—so to speak—of the dependence of others. Her wards are one frightful old woman and a listless young girl. The first demands an increasing amount of time and quite intimate care, from a woman whose calendar is her backbone and who prefers the uncorporeal tidiness that her generation has been the first to claim. The other moves in and hangs around the house, a vacant, stupid girl with no interest and nothing to do. She invites in equally cast-out characters, some criminal, who break things and create general filth.

These novels do not in any way redeem their subjects. The old woman is unlovable, the young girl despicable. Jane takes responsibility not because she wants to, or even because she does not dare not to. She gets caught up in situations and deals with them without deciding to. There is nothing good about it, or humane. No moments of serenity where souls mingle or a glimpse of meaning sparks forth. Neither do we read explanations or analysis. Lessing delivers a raw chunk of prose with no special densities or colorings. This is of course "style," as with any other kind of prose; but at the same time Lessing seems to have totally withdrawn from aesthetic beauty. One might say that she has moved her frontier from the healing accuracy of expression to the mirror image of life's unorganized misery. She no longer trusts the make-believe of language and has let her words slip from the manners of realism to the grey zone of reality, where definition solves nothing.

Returning from pseudonyms back to her famous name, she has kept up the no-nonsense attitude of the *Diaries*. *The fifth child* is a perhaps even more "artless" product, which also stresses what this is all about: It is a story about a young couple living their plans. They have good educations and fulfilling professions, they buy a large Victorian house with a wild (though planned) garden, share the right, humane opinions, and efficiently keep together a vast amount of family, relatives, and friends through holidays and festivities. They are not rich, and their entertainment has the unpresuming style of an Italian farmhouse kitchen, where everybody sits together peeling onions. Good, generous, and undemanding attitudes should be proof against malicious destiny—that is a promise in the undercurrent of left and green rhetoric. But when the fifth child is born, all the plans and promises are upset. He is a freak, ugly and murderous though still in diapers. His very existence ruins the gentle tone of the house, which faces an exodus first of friends, then of relations and family. Lessing uses a cool, onion-peeling voice to tell the story. There is no emotional stress to register opinions on either side. The naïveté of a family who believe they have the right to plan disaster out of their lives is told without irony. The fifth child gets neither distaste nor compassion from his creator. At the end of the novel the child's mother sits alone with him. Everything else has been lost because of him. But she seems content with the collapse of things, without feeling or reflection, as though the rudeness of reality has somehow freed her. The author of the novel sits equally heavily, going right to the very dregs of things.

The dregs have been Lessing's point of friction since she gave up accuracy and definition: Everything outside human planning, which has been mentally nonexistent in the welfare state and now—without plan but through amassment—is staging its re-invasion of a society that considers itself legitimately lucky. One might argue that the "badness" of her style is so acquired to fit the theme, that she has really not given up the author's friction with language. But insofar as she directs the reader's gaze to frontiers and breaking-points in her narrative, she now fixes it where intellect, organization, and planning face a growing mass of nameless matter, the one we thought was ruled out of modern life for good.

In Rebecca West's trilogy about the Aubrey family (*The Fountain Overflows, This Real Night, Cousin Rosamund*), there is one person whom nature has equipped specifically with means to surpass friction. She doesn't understand better than to give in to the temptation, nobody tries in earnest to influence her, and eventually she dries out and withers from her lack of sound conflict with life. Indeed, the absence of friction training in youth means that she doesn't know how to avoid getting stuck later on. Instead of developing, she hardens into a philistine, whose erstwhile pretentious artist's image is quite discarded and forgotten.

Her name is Cordelia. She is the oldest sister and the prettiest, and the one with the best ear for music in a musical family. Looking sweet with her fiddle and from an early age very swift with the bow, she wins the humbug victories of the *wunderkind* and never advances from there. She gets encouragement from teachers with starry eyes and talent for intimate friendship. The uncritical audience of the pantomime loves the young girl with red curls and amazing technical skills. The only ones who would like to bury her violin are her family and their friends, all of whom have—no, not just taste, but a sense for what is genuine and real, and *earned*. To describe Cordelia's music between themselves they use "friction-words" like *oily*, *slithery, treacle.* They also show cool insight about the fact that she cannot be reformed. Her entire personality is set to succeed, and success in itself is an end—whatever the quality and however she comes by it.

There is no mercy about the fact that she is singled out in a set where everybody else does it the right way, or rather the hard way. They are poor and proud. Things possessed must be exquisite (even though they are mostly cheap or found and gathered) or not at all. Food and clothing are insignificant, one must stay detached from such matters, and incidentally it's no problem, since they lack allure when one directs one's energies toward the sublime. Rebecca West does not create a family of monsters. They have their share of human weakness, but they are extreme snobs about art, work, and character—each must be pursued like truth itself. They *are* truth itself—in different incarnations. The children of the family—except Cordelia—need no basic training of character. They seem born into a gentle consensus and wear themselves out to preserve it in a world of futility and nonsense. With ears sensitive to false notes, or notes slunk through in a cute manner, they choose people by their hearing, sometimes demonizing people to "geniuses" just because they like them. There is nothing fine or fancy about their lives except the rule of earnest pursuit.

When standards of this consensus clash with the alien—vulgarity, grossness, or evil—odd things occur. A murderess can be all right it she is forced to murder by ugly facts. However, someone who is just a moral or aesthetic misfit will not be excused, but excluded: Cousin Rosamund is not outspoken about the bad relations between her sisterly mother and her offensive father. She doesn't seem to notice them very much. But when her desire for peace meets with the disharmonious facts of home life, she has to "shed energy" somehow. She is a placid girl and therefore subconsciously brings forth poltergeists, which wreck the house, until the man gets the message.

Another image of friction surrounds the deathbed of Mrs. Aubrey. After the labor, the waiting, the suffering, the advancing, she pushes through the hymen of final knowledge and shoots "like an arrow" into space. West builds her entire trilogy upon the extremely slow process of insight. The day-to-day life of the Aubreys seems a series of mutual confirmation. But the summarizing that inevitably takes place over the years also undermines the consensus. There is wear and tear. Sometimes the friction comes to surface, as in the earlier examples. But as an overall phenomenon it delivers a surprise at the very end of the story. The narrator—one of the Aubrey sisters—has through the entire span of events kept total faith with the family enclave and its household ideas. The first rift appears when she realizes that she and her sister are not the same kind of pianist. They have had equal instruction and opportunity and practiced equally, and the general presupposition seems to be that they are equally good. The reader is effectively lured into believing this as well. Insight about the other sister's vast talent and the narrator's relative mediocrity is not forced upon her (and us) by truth-loving strangers. The narrator herself works her way to the pain of knowledge, and when she is mature enough to take it in, it doesn't hurt so much.

Fairly late in life she does manage to fall in love outside the family circle. For the first time her affection is directed toward something strange, male, and sexual. And once again there is a breakthrough, though prepared for through anticipation on the reader's part—the family web has begun to seem all too close, almost sickening in its proud privacy. Then, without warning, it is shockingly shredded. The illusion of hermetic constancy, into which the reader has been lulled for three whole books, suddenly is abandoned by the narrator with no sense of loss. This was where she was going, but it took all the beauty and misery of before to get there. Insight comes as a surprise only to people on the outside. The one with the experience has known the narrow stages on the way.

Wrestling

Jacob's wrestle with God in Genesis is a basic scene of friction, not possible to surpass in its simple concentration. Reading it over you taste the essence of identity. Like everything in this book, the event in itself is told with a crude sort of ease. Jacob wrestles with "a man" all through the night. When dawn comes and the man realizes that he cannot conquer his adversary, he strikes a blow across his hip and dislocates it. Then he asks to be released. Jacob refuses to let the man go until he has blessed him. This incident is told in few and simple words, rich in symbolic value although not artfully so. The effect is frugal and laden at the same time. As a reader you sense the duration and endurance of a night-long wrestle, and although you cover the text in a couple of minutes, you experience a similar—or endless—duration in your attempt to grasp them. Jacob will not let go of his adversary even after his hip has been broken. The blow and the blessing become each other's prerequisites. They move tightly against each other, like millstones, and the product of the grind is indeed fine as flour: *identity*. Having persevered until this painful blessing, Jacob has his token name—Israel—bestowed upon him. To this day very few words or names create a stronger image of identity than Israel, and looking back along the lines of it you can easily perceive a continuation of that night's wrestle.

Earlier, I mentioned appeal as a smooth and effortless shortcut to identity. The naming of Jacob is the kernel of a direct opposite—a faceless, exploring ordeal, which eventually bears fruit in one hard-earned *name*.

Isak Dinesen frequently refers to this scene in her allusions to the value of hardship—or rather to the way that hardship created her identity. She does not resort to the banality of claiming that she was constantly aware of struggling for a sublime cause. Hardship was simply hardship and banality as such. But what seemed futile and broken in her day-to-day existence turned out to be "wholeness" and purpose when she put it through the process of her fiction, which became intertwined with her life image. Her struggle—like Jacob's—was the travesty of embrace. She fought with her own self to turn feebleness into strength. Some of it might have been cosmetic, but still. She tried to force an unwilling nature into bearing fruit and there found a gently constant opposition, nimbly adjusting its strength to her strivings. Reality never gave in to her ambition, but the way she tried and pretended chiseled out her identity.

You are never wholly aware of what happens to you in life or of the long-range significance of events. Isak Dinesen evaluates her wrestle with hindsight and finds that the loss of worldly goods—as well as of prospects, love, health, position, and youth—still left her with one cool jewel of decisive importance to herself: identity.

Her image of how life summarizes into purpose is the story of the stork-drawing in *Out of Africa*: A man on his errand strays and stumbles, falls and falters. The whole performance seems idiotic, like a beetle trying to find his way on a Kelim carpet, until you use a pencil and find that the picture of a stork could be drawn along these erratic wanderings. The very unimportance of a stork-drawing as such is one of the author's main points, of course. Abstract importance is apolitical matter. Of importance to *you* should not be market items but what you yourself truly value and what enriches your identity.

In art—especially in art—the experience of friction may lead to outwardly insignificant results, which suddenly illuminate by the thoughtfulness and thoroughness through which you arrived at them. Things of importance are the liquid you manage to wring from an almost dry cloth—as opposed to the generous gush of nonsense that will so readily engulf you.

Dinesen made her appearance as a writer late in life and had the benefit of retrospect when she considered these matters. According to the letters she wrote while in the midst of failure and frustration, she could not be the stoic at the time; at best she could merely pretend. You can also argue that only success in the end gave her the ability to value friction and failure, and that common-sensical wordings, such as the one about how suffering will ennoble you, are all too close at hand. A noble or at least trained mind will certainly help you bear, or make use of, friction. It is only if you can savor it that wrestling with something superhuman would leave you broken and blessed with the identity from this encounter.

So you need to have—in some sense—a religious mind to be able to "enjoy" friction. There is pain and weariness, and even quite amusing hopelessness, involved. Friction means exposure to life and lifelong equilibrium between attempt and failure. The duration of Jacob's wrestle with God is the point of that story. And the duration of Dinesen's combined struggles is the point of hers. Success and failure, loss and gain, are intertwined in these long conflicts. Gradually the symbols of either relax their significance and melt into one. The raising of energy and the shedding of energy are not means but in fact the very End.

References

Dinesen, Isak. *Out of Africa*. New York: Modern Library, 1992

Lessing, Doris. *Jane Somer's Diaries*. London, 1984.

———. *The Fith Child*. London, 1989.

West, Rebecca. *The Fountain Overflows*. London, 1957.

———. *This Real Night*. London, 1984.

———. *Cousin Rosamund*. London, 1985.

9

Why Things Don't Happen as Planned

Jon Elster

This chapter is about thwarted plans and frustrated expectations. The experience that things don't always work out as planned is a common one. However, instead of just citing Murphy's law or the inherent malevolence of the universe, we need to understand the structure of frustration. Sometimes, things go wrong simply because we make a mistake, such as stepping on the accelerator instead of the brake. I am concerned here with more systematic, recurrent sources of frustration. Some of these are located within the individual; and others, in the interaction among different individuals.

I begin with two complementary phenomena, *weakness of will* and *excess of will*. The former and better known of these mechanisms has the following structure. We want to do *x*, for good reasons. We also want to do *y*, for good reasons. However, *x* and *y* are mutually exclusive. All things considered, we believe the reasons for doing *x* are stronger than the reasons for doing *y*. And yet we do *y*. Weight Watchers and Alcoholics Anonymous owe their existence to this mechanism. It could also be illustrated by cases in which *x* is to quit gambling, stop smoking, start jogging, or begin saving for one's old age. These are all instances in which short-term interest as *y* is opposed to long-term interest as *x*. In other instances, *y* could embody self-interest and *x* some more altruistic concern.

But *x* and *y* may also be turned around. Individuals with rigid and compulsive personalities are often unable to give themselves a break. They may well believe that on a given occasion, all things considered, short-term self-indulgence is justified and even recommended, and yet be unable to deviate from the rules they have set for themselves. This is not the phenomenon of excess of will that I consider later. Rather, these rigid characters are weak-willed, prisoners of their self-control techniques. The best perspective on such phenomena is, I believe, a neo-Freudian one. The ego is constantly engaged in a two-front war, between the impulses of the id on the one hand and the impulse-control mechanisms of the superego on the other. So the ego has to fight against both the primary problem and the secondary problems that arise in trying to cope with the primary one.

To connect this analysis to the idea of frustrated plans I shall focus on the special case in which x and y represent, respectively, long-term and short-term interest. Typically, three temporal moments are involved. At time t_o, we know that at time t_1 we shall have the choice between getting a small reward at time t_1 and a larger reward at time t_2. The weakness-of-will scenario, then, is as follows. At t_o, we decide that we would rather have the later and larger reward than the early, smaller one. As t_1 approaches, however, there comes a moment when preference reversal occurs. The imminent availability of the earlier reward clouds our judgment, preventing us from sticking to the decision we made in a cool and considered moment. What happens in a case like this is that our reasons for choosing y become stronger than the reasons for choosing x—not stronger *qua* reasons but stronger *qua* sheer psychic turbulence.

The idea of excessive will encompasses a very different set of phenomena, which can be summarized as "willing what cannot be willed." It is notoriously the case that certain states are very recalcitrant to attempts to bring them about intentionally. They are essentially by-products of actions undertaken for other ends. Paradigmatic examples include the efforts to overcome impotence, insomnia, calculatedness, unpleasant memories, and stuttering by concentrating hard on the desired state—lust, sleep, spontaneity, forgetfulness, or uncluttered speech. The very concentration interferes with the state one is trying to bring about, which essentially requires a relaxed, un-self-conscious attitude.

On reflection, one can perceive this mechanism everywhere. It explains, I think, the notorious failure of self-help books. (Of course, many such books succeed in realizing the goal of the author, which is to bring in money. What I have in mind, though, are the goals of the readers.) Self-esteem is essentially a by-product: It supervenes from other activities that are pursued for reasons other than that of providing self-esteem. The same mechanism also explains, or so I believe, the failure of certain social movements, whose members join them less for the sake of the cause than for the impact that joining will have on their own character formation, neglecting to acknowledge that this impact has to be mediated by a belief in the cause. Character, in fact, is essentially a by-product.

Many failures of rationality are due to the choice of inappropriate means to a given end. With excess of will, the problem runs deeper. The mistake here is the belief that the desired state is such that it can be realized by means-ends rationality. Instead of talking of excess of will, one might well talk about *hyperrationality*. But that phrase could also be taken in a different sense, to refer to the desire to find reasons or give reasons for decisions that essentially cannot be rationalized. An example will indicate what I have in mind. Suppose a surgeon is brought to a disaster area where many people are in urgent need of surgery. Assume, for simplicity's sake, that he has to decide which of two patients to operate on first. To make a rational decision, he has to assess the relative urgency of the two cases, which he does by means of the appropriate examinations and tests. While he is doing so, however, first one and then the other patient dies. The surgeon could have saved one of them had he operated without undergoing the ritual of rationality.

This example is probably far-fetched. But there are other, very real cases in which rationality proves to be self-defeating. In child custody and adoption decisions, the rule nowadays is that the choice of custodial or adopting parent should promote the interests of the child. Courts and adoption agencies must therefore compare the parents or adopting couples with respect to their ability to raise the child. In disputed custody cases, especially, this can be a very time-consuming process. Typically, the child suffers a lot, not least because he or she is used as a pawn in the quarrels between the parents. On average, the child would have fared better if the courts had not sought to promote the interests of the child but, instead, had used some swift, mechanical decision rule such as the principle of maternal presumption or even a lottery. In adoption cases, the problem is compounded by the fact that as children grow older, they become less adoptable. When parents rather than children are the bottleneck (as with the adoption of black children in the United States), the insistence on waiting until the optimal parents come up may have the result that no parents are found at all. The best, in such cases, is the enemy of the good. Insisting on finding the decision that would have been optimal if found instantaneously and costlessly is pointless when the search for this solution is in fact both time-consuming and costly.

Both varieties of hyperrationality—excess of will, and the search for the optimum that neglects the costs of searching—are instances of *self-defeating plans*. But they also differ. The latter problem can be solved, once it is recognized. However, as I later explain in more detail, recognition is often made difficult by the deep need that human beings seem to have for being able to give sufficient reasons for what they do. The former seems more untractable. In some cases, to be sure, one can use indirect strategies. Instead of trying to will sleep to come, one can take a sleeping pill. Instead of trying to induce religious faith by sheer will, one can engage in religious acts that have the predictable result of inducing faith, while simultaneously inducing forgetfulness about the process itself. This was the strategy proposed in Pascal's wager. But not all the states that I characterized as essentially by-products can be attained by indirect strategies. Once one has eaten of the apple, there is no way back to innocence and spontaneity.

Our minds play many tricks on us. In some cases, the result is peace of mind in what would otherwise have been an uncomfortable situation. This outcome is illustrated in the fable about the fox and the sour grapes. For the fox, wanting the grapes and knowing that he couldn't get them induced an unpleasant state of "cognitive dissonance." By being able to tell himself that they weren't worth having anyway, he made his life easier. But consider another case. I want a promotion, but there are few chances that I will get it. In that case, I can imitate the fox and tell myself that the higher position is really too burdensome to be worth the extra pay. But I can also refuse to believe the evidence and persist in this unjustified belief that promotion is just around the corner. This will, for a while, also bring my beliefs and desires in harmony with one another. However, the mistaken factual belief can also get me into trouble, for instance, if I buy a new house on the strength of my expectations.

This is not to say that we would be better off without our passions. Without them, we might as well be dead. We would be better off if we could have the motivating power of the passions without their distorting power, but that may not be a feasible combination. The motivation to make plans and to follow them up is positively correlated with the strength of our passions. The ability to follow them up by choosing the appropriate means is negatively correlated. In particular, the passions tend to make us overestimate our own abilities. By and large, the only individuals with a correct appreciation of their own capacities are the clinically depressed. Imagine that achievements are ranked on a scale from 0 to 10. If I believe I will achieve 9, I shall in fact achieve only 6. If I aim at 7, I shall realize 5. It is only by limiting my aim to 3 that I can actually achieve what I set out to do. My plan, in the last case, suffers no frustration, but at a price I might not want to pay (assuming I have the choice, which I don't).

When preferences adjust to circumstances, the result is, as I said, to give us some peace of mind. But the opposite mechanism may also be observed. Sometimes, we want what we cannot have simply because we cannot have it, and the moment we get it we no longer want it. Forbidden fruit tastes best, and the grass is always greener on the other side of the fence. When we live in New York, we dream only of living in Paris, and vice versa. Or consider two modern and premodern conceptions of love. The premodern conception is encapsulated in Hermione's question in *Andromaque*: "Je t'aimais inconstant, qu'aurais-je fait fidèle?" or, I loved you while you were inconstant, what would I not have done had you been faithful? The implicit answer is clearly that she would then have loved him even more. The modern conception is described by Stendhal and Proust. Love relationships are like a see-saw: When one is up, the other is down. There is no such thing as happy, consummated love.

Nor is there such a thing as successfully consummated hate. For the person who defines himself by his opposition to a person or a doctrine—the fanatical antiroyalist, anticommunist, or atheist—the worst that could happen would be the actual success of his destructive efforts. John Donne, in *The Prohibition*, said it this way:

> Take heed of hating me,
> Or too much triumph in the victory
> Not that I shall be mine own officer,
> And hate with hate again retaliate;
> But thou wilt lose the style of conqueror,
> If I, thy conquest, perish by thy hate.
> Then, lest my being nothing lessen thee,
> If thou hate me, take heed of hating me.

So far I have looked at some sources of frustrated plans that are internal to the individual. In this second half of the chapter, I shall consider frustration due to the structure of interaction among individuals. To some extent, as we shall see, these frustrations can also be described as caused by individual failures to form rational beliefs—that is, beliefs about what others are likely to do.

Sartre coined the term *counterfinality* to mean a certain type of unintended consequence. In his example, we are asked to consider the process of erosion in the Chinese countryside. For each individual peasant, deforestation is a rational strategy, given his pressing need for more land. However, when all peasants behave in this way, erosion sets in, and they all end up with less land than they had initially. The causal mechanism underlying this story is one of *externalities*. By cutting down the trees on *his plot*, each peasant raises the probability of erosion on *all plots* by a small amount. His utility calculus tells him that the risk is worth it. However, universal deforestation raises the probability of erosion to a certainty. In the collective utility calculus, it was not worth it.

I have stated the peasant's dilemma as a Prisoner's Dilemma. But in a way this is misleading. In standard Prisoner's Dilemma situations such as pollution or congested roads, all actors are fully aware of the causal structure of the situation and know that others are so aware. *Nobody is surprised* when they find the park full of litter or the roads blocked by cars, for this is the predictable outcome of others behaving as they do. But when peasants lose their land, they are surprised. Initially at least, they will tend to ascribe the outcome to bad weather or the anger of the gods. Only after a while, if ever, will they come to suspect that the deforestation is the result of their own behavior. And even then they may not understand the causal structures involved. Perhaps each peasant comes to believe that the erosion on his plot is fully explained by the deforestation on his plot. He may then try to remedy the situation by stopping the deforestation or planting new trees, and be frustrated again when he finds that it doesn't help.

Rather than pursuing this speculation, let me turn to the better-understood and better-documented mechanism underlying the *cobweb cycle* (so called because if illustrated in a demand-and-supply diagram the mechanism looks somewhat like a cobweb). I shall illustrate the mechanism by an example from farming, but it is much more widespread. In Norway, for instance, overinvestment in the shipbuilding industries in the 1970s and in the fisheries in the 1980s follows exactly the same logic. And innumerable examples from other countries could be cited.

Suppose that in year 1 farmers (in a competitive market) are making up their mind about how much to produce for year 2. To do so, they must estimate what prices will be in year 2. A natural assumption is that prices will remain constant. Planning on that assumption, the farmer will market a certain volume in year 2. Now suppose that year 1 prices were unusually low. With capitalist farmers, this will induce a low level of production for year 2. With lower prices, the fact that capital and labor have decreasing marginal productivity makes it necessary to produce less in order to equalize price and marginal cost. With family farms, a high level of production will be observed. Since the family cannot fire any of its members, they now have to produce more to earn a decent income.

In both cases, the farmers will be surprised. In an economy of capitalist farmers, they will be pleasantly surprised when they observe that they get very high prices at the low volumes produced by all. But in an economy of family farmers, the surprise will be an unwelcome one, as they observe that at the large volumes produced by all, prices have fallen to even lower levels. This mechanism was observed in the Nordic countries during the Great Depression and ultimately led to regulation of agriculture. In the following pages I shall leave the second mechanism aside and follow only the development of the capitalist case.

In year 2, then, capitalist farmers will produce little and get a good price. My assumption is that, when planning for year 3, they assume that prices will remain high. The logic of equalizing price and marginal cost will then induce them to plan a large volume for year 3. However, when all act on that plan, the large volume thrown on the market will drive prices down. In theory, this cycle can go on forever. In one possible scenario, there will be explosive cycles, in which the below-average prices that obtain in alternate years will fall more and more and the above-average prices that obtain in intermediate years will rise more and more. In another scenario, the cycles will be damped, so that the difference between below-average and above-average prices is steadily reduced. In the latter case, prices will eventually converge to the equilibrium level defined by the intersection of supply and demand curves. There will be no more surprises, no more frustrated plans.

But consider the former case. As prices oscillate in ever-widening cycles, it is not unreasonable to assume that the farmers eventually come to understand that this outcome is their own doing. Each farmer will come to see that a year of low prices tends to be followed by a year of high prices. He will therefore adopt a new principle to generate his expectations: Instead of assuming that next year will be as this year, he will come to think that next year will show the opposite pattern. However, his frustration will not end here. For when all farmers form their expectations on this new pattern, they will all be proven wrong. In a year of low prices, each farmer expects that prices will be high next year. As a consequence, he plans for a big volume. Next year, however, when all farmers dump their large crops on the market, prices will not reach the expected levels. Even if the farmers eventually understand what is happening and adjust accordingly, this will only reproduce the problem at a new level.

The root of the problem is that each farmer believes himself to be one step ahead of everyone else. The initial assumption, that prices will remain constant, amounts implicitly to an assumption that all other farmers are like automata that mechanically do the same thing year in and year out, so that he is the only one to adjust to market conditions. After he (together with the others) learns that others are in fact adjusting to the market, he now (along with all the others) acts on the tacit assumption that he is the only one to have understood this fact. But *each farmer cannot be the only one* to have seen through the mechanism.

Similar phenomena can arise in politics. Often, voters make their decisions on the basis of previous opinion polls. Suppose that the polls predict that the Labour Party will win by an absolute majority of 55 percent. Some voters on the left wing of the Labour Party will then switch their vote to the party immediately at the left, to send a message to the Labour Party to change its policies in that direction. But if more than 5 percent of the voters act in this way, the Labour Party might lose the election, contrary to the expectations of these voters. Each of them has tacitly believed that all other voters would behave as they had told the pollsters they would, so that the Labour majority would be safe from defections.

In these cases the frustrated expectations are due to the failure of the farmers or voters to understand the strategic nature of the situation. Once they come to understand that it would be irrational to believe that others are less rational than themselves, they can form rational expectations with the self-fulfilling property that when all act on them they will be exactly fulfilled. The equilibrium prices defined by these expectations have the property that when all but one of the farmers make their plans on the basis of them, the remaining farmer has no incentive to do otherwise. In fact, this is the general definition of what is called Nash equilibrium in economics: a set of strategies, one for each actor, with the property that each strategy is the optimal response to the others. In equilibrium, nobody has an incentive to deviate.

However, the existence of an equilibrium (and one usually exists) is not enough to ensure the nonfrustration of expectations and plans. In the first place, the actors may not know enough about each other to converge to the equilibrium (assuming there is only one). Only if the actors have a dominant strategy, as in the Prisoner's Dilemma, will they be able to figure out what to do without figuring out first what others are going to do. In the absence of a dominant strategy, they cannot determine the equilibrium without knowledge of the preferences of the other actors, and even with this knowledge they may hesitate to use their equilibrium strategy if they are unsure whether others have enough knowledge to do the same.

But there is an even more serious problem. Sometimes, there is more than one equilibrium. This need not create any coordination difficulties, but it often does. If one equilibrium is better for everybody than all the others, it will obviously be chosen (if the knowledge requirements are fulfilled). But often one equilibrium is better for some, and another better for others. If only two person are involved, considerations of bargaining power may help us to predict the outcome. But in large market situations with many equilibria, the outcome is inherently indeterminate.

Now, indeterminacy need not yield frustration. If the actors understand that the equilibrium outcome cannot be uniquely determined even under assumptions of full rationality and perfect knowledge, they will have to adopt an imperfect decision heuristically, fully knowing that after the fact they may come to see that they could have done better. Investment decisions by firms are, for instance, very largely indeterminate in the sense just indicated. A given firm will not be able to calculate how much other firms will invest on the basis of mutual rationality and self-fulfilling expectations. Instead of optimizing, the firm can then try do so as well as it can under the circumstances, for instance by following a rule of thumb such as "invest the same proportion of profit as last year, unless the rate of return on capital falls below 6 percent," with some additional rule about what to do in the latter eventuality. Although there is no reason to believe that this is a optimal response to what other firms will do, no other policy is demonstrably better either. After the fact, the managers may be *disappointed* when they see that they could have done better; but as they will know that they could not have known, they will not be surprised.

Such sagacity is probably rare. More frequently, managers persuade themselves not only that their policy is no worse than any other but that it is optimal in light of what others can be expected to do. Human beings, as I said, have a very strong need to act in terms of sufficient reasons. This deep-seated need for rationality is in fact an important source of irrational behavior. It is also an important source of frustrated plans. Any mechanism that makes us form expectations when there are no proper grounds for doing so will also induce frustration.

References

Although specific citations are not provided, this chapter has drawn extensively on some of my earlier writings. Here are some brief references. For a discussion of counterfinality, see chapter 5 of *Logic and Society* (Chichester: Wiley 1978). Regarding weakness of will, see chapter 2 of *Ulysses and the Sirens*, rev. ed. (Cambridge: Cambridge University Press, 1984). Regarding states that are essentially by-products, see chapter 2 of *Sour Grapes* (Cambridge: Cambridge University Press, 1983). Regarding the indeterminacy of rational action, see chapter 1 of *Solomonic Judgements* (Cambridge: Cambridge University Press, 1989). Child custody dilemmas are discussed in chapter 3 of this last book. Regarding wishful thinking and its consequences, see chapter 4 of *Sour Grapes*. Regarding the idea of adaptive and counteradaptive preferences, see chapter 3 of *Sour Grapes*. Regarding the idea of inherently contradictory goals, see chapter 4 of *Logic and Society* and chapter 4 of *Ulysses and the Sirens*. Needless to say, many of the ideas developed in these books owe much to the work of other authors. A compact set of references will be found in the bibliographicla essay at the end of my *Nuts and Bolts for the Social Sciences*, Cambridge University Press.

10

Six Poèmes en Prose

Klaus Rifbjerg

Naval hero Tordenskjold died in a sword duel. He spiked the Swedish cannon with cunning and took Marstrand. Tordenskjold won the battle of Dynekil. When still a boy, his father forced him to wear a pair of lederhosen. Peder Wessel—as he was called before he became a hero—always wore out the seat of his trousers. A stop had now been put to that. There were other children in the family and trousers cost money. But this did not stop the lad. He sat athwart a grindstone and asked a group of other scamps to turn it. In the end the inevitable occurred. He wore a hole in his trousers. The way his father looked when the boy presented him with hole-ridden lederhosen has not been recorded. Nor the way his mother looked at the time. But Peder Wessel from then on wore ordinary clothes right up to the time he became a hero. Then he appeared in uniform and adopted his new name.

The steam locomotive is standing on the rails. The stoker has been working for a long while now with the fire under the giant boiler. To make sure that everything is in order, the engine driver arrives early and inspects the valves and the manometer and the long iron lever that allows you to transfer compressed steam from the boiler to the pistons. In the carriages— right behind the tender—the passengers are collecting. There is a great bustle, and the platform is filled with a lively commerce. Sausages and ice cream are handed to those on board, a child is crying, women lift up their skirts as they ascend the high step, the engine driver shows where they have to sit. The moment is approaching; it has not quite arrived. There is time to wipe the sweat off one's brow. Time to consider that only during travel is it permitted to use the toilets. Time to exceed the timetable or leave it be. Time to time. Until time is up and the steam gushes through the ingeniously constructed system with a hiss and the pistons engage and the power is transferred to the drive-shafts and so to the wheels. Which stay still. Until they slip (under pressure) and turn (without pulling) at a rapid rate. At a rapid rate! Without pulling! Until they pull and the train moves off.

"The old in-and-out" the English call it. But the phenomenon has been described in numerous ways, and other names have been given to it throughout the ages. You only have to think of anatomy and poetry! But when it comes to the crunch, it is simply a question of an in-and-out movement. It can be rapid or slow, it can be a question of movements so small that they are hardly registered, in reality standing still. But it is disingenuous to talk of standing still. It is not standing still in the serious sense of the words. The expectation of the ultimate movement, the last and decisive one, is too great. No one forgets—at any rate, no one who has tried it—that electric thrill when the muscles of the vagina rub against the underside of the glans. Out and in. In and out. That is a matter of secondary importance. The aim is reproduction. The sacred end justifies the means. *Sacred* is the word.

"Swing it, professor!"—and everyone thought it was the Swedish star Alice Babs, who created the sensation. People stood in a queue in dead, gloomy Copenhagen, rationed to pieces as it was and occupied by the Germans, in order to catch a glimpse of the *girl from the other side*. She was lovely, sang, would hardly be struck by a stray bullet or be executed for illegal activities in a backyard. She could travel home whenever she liked, home to glittering Stockholm with its neon advertisments, chewing gum, and American films showing at the cinemas. There he sat at the grand piano. There she stood. Now she turned her head and smiled. Now there was a pause. Now a foot was beating out the rhythm. Now nothing was happening. Now something was happening. But it was not only in the dizziness of the lifted whiskers, or in the ritardando of the raised foot, or in the sudden flash of the smile that the real sensation lay. It was the longing to be able to dance oneself which hid itself in her throat and in the rhythmic retardation of her hands; it was the capitulation, the dream of freedom "in the darkness before a new dawn."

The huntsman puts the shotgun to his cheek and aims. The huntsman slams the shotgun against his cheek and doesn't aim. The bird appears from nowhere. The shot and the cartridge sit compact and indolent behind the charge. The natural, absurd confusion of the moment is concentrated into decision, it is quite intuitive. Of course the bird was there first. The huntsman slams the shotgun up to his cheek and follows the bird with both eyes open. He glues himself to it and overtakes it with a small, silent pop. That's how it is when you *glue* your eye to a bird on the wing and overtake it. You get a popping sound. It can be mistaken for the shot that immediately follows. But it is there, the pop, ask anyone. The shot wakes up to the kick of the gunpowder, which is activated by the trigger, the pressure from the finger, etc., etc. The barrel shapes the swarm of shot and gives it extra speed. And Bob's your uncle: a dead bird.

"He who has patience will win through," says an old proverb. Death has invented it and spread it abroad. He knows the rhythm of the saw and has a grip on matters right from the start. It is a matter of not being too hasty, not cutting at too steep an angle. "There is no haste for he who believes," he says to himself. "Sawyer, sawyer, saw some wood. . . ." He knows all the rhymes and the old songs. Pull and scrape, heave and haul. But never push. Gradually, little by little, the sawdust piles up in his footprints, trickles from his ears, runs from his arse. Death never comes sweating. He is quite content to lean back and take a rest. "More haste, less speed," he says to himself. So one day he is right down to the bone, the marrow and the root. And it is here that his reward lies. For it is not merely a question of another victim, another death, another fallen hero or coward. It is the moment *before the fall*, the ultimate and definite one that he has been waiting for, that ecstatic moment when the decision has been taken but not yet carried out, the orgiastic after-beat where the mighty trunk dances for one second in all its absurd and untamable height, before the whole caboodle collapses, tediously falling to the ground with a theatrical crashing and rending.

[Translated from the Danish by Eric Dickens and Nordal Åkerman.]

PART SIX

Structuring the Human Space

11

The Desire for Order

Joanne Finkelstein

The Modern World

The process of becoming modern has been the conquest of the mysterious and the unpredictable. We humans have declared ourselves to be the supremely rational animal capable of ordering knowledge, evaluating its significance, and applying it in the most efficient way. From this manner of thinking, the modern era has come to associate the achievements of the rational scientific method with the measure of civilization. As such, the future of the modern world is represented as a matter of technical application: A better world is depicted as a more controlled, less ambiguous, less confusing, less contested one. The end result is a world thought to be malleable enough to be perfected. A better world, according to this view, becomes merely a matter of technical virtuosity.

The early-nineteenth-century social reformers, including Auguste Comte, asserted that society could be an orderly creation shaped by human desire and brought into being by human will, much like the great machinery of industrialization. This vision, in one guise or another, has endured as a defining characteristic of the modern. Society as a system or machine has been a widely used metaphor that gives emphasis to purposive, functional, and reason-guided actions on the basis that these represent the best way to achieve civilization. There has been widespread acceptance of the power of rationality and the pragmatics of economy and effectiveness as the principles by which to determine how we should live. We have come to regard technology as the best means of freeing us from tricks of nature, chance, and metaphysical circumstance. We have created a systematic bureaucracy on the basis that repetition and uniformity produce a better culture than the assumed alternative, namely, the exercise of arbitrary and willful power. We have come to accept a view of the modern world as being knowable and controllable. It is as if the world were more and more transparent, with its inner properties and functions coming more sharply into view.

But, increasingly, cracks in the veneer are being exposed, and the way to a civilized future seems less straightforward than is implied by this rationalistic perspective. Without doubt, the twentieth century has been an era of vast invention and grand acts. It has also been a century of world war and continuous parochial battle; of massive population explosion, migration, and extermination. It has been a century of revolution, repression, and diverse intellectual expression. Ideas have been produced in abundance, as have goods and services. Technological inventions have irreversibly altered the topography of the planet, leaving little doubt that this has been a century of conquests—over nature, culture, and the human imagination. However, the modernist belief in the possibility of an orderly future and the confidence expressed in the scientific enterprise have been increasingly tempered by the burgeoning numbers of environmental crises, economic failures, and unanticipated political consequences that stain the progressive ideals of modern thought.

In the late twentieth century we have seen too much of political cynicism, human cruelty, and economic injustice to maintain a firm belief in the civilizing force and inexorable progress of the modern society. The romanticized manner of thinking that has undergirded instrumental reason and the belief in a cost-effective form of social development has faltered. The catastrophic events of the century—the Gulags, Hiroshima, Nazism, and the creation of a Third World, as well as the daily encounters we individuals have with a blind, unresponsive bureaucracy, the senseless competitiveness of rampant consumerism and ostentatious display, the interpersonal violence of exchange in business, the displays of hostility when people share public space on crowded roads or wait in service queues—have significantly undermined a belief in a rational calculus that can bring about the best possible world.

We have come to know our era as one in which ideas of progress have a much weakened force. The ideal of an autonomous and egalitarian society developing from the rational application of technology and knowledge has not been realized, and, contrary to expectation, it is becoming more difficult to accept the simple trajectory that from observable, universal rules there will emerge an orderly, stable future in which emancipated, fully realized human beings will live happily ever after.

Making sense of the modern now requires a greater willingness to accept disappointment, chaos, and complexity. We have arrived at the moment when the ideals of the modern—to control the natural world and remake it in the name of civilization and emancipation—have turned out to be something of a stranglehold. Indeed, two pinnacles of the modern, science and the city, are emerging as massive failures of modernist ideals. In the case of modern science, its technological accomplishments are beginning to reveal unanticipated consequences such as environmental degradation and overproduction of unnecessary goods. In the case of the modern megalopolis, its competing cultures of commerce, community, and cosmopolitanism are being corroded by the growth of corporate power and the amassing of vast wealth; this has divided the city into artificial neighborhoods of activity, creating the opportunity for graft and directing governmental policies away from egalitarian principles toward specific interest groups.

The increasingly visible problems of the modern have produced a new set of issues that emerges from the hidden dimensions of the social enterprise and its previously unimaginable and unanticipated consequences. It is becoming clear that the modern world is not just about high technology, scientific advances and cultural proliferation. It is not just about more music, art, and literature, it is also about destructive technology, exploitative art, repressive power, and unsatisfying culture. The social experience is more disorderly and fragmentary, more contested by irreconcilable interests, than was ever alluded to in the classical evolutionary histories of the recent past.

The period of human history known as the modern is an arbitrarily dated epoch. For most purposes, it is useful to define this period conceptually as one produced by a number of sociohistorical shifts and convergences involving the rapid technological expansion that accompanied scientific discovery and industrialization: the growth of the city and urbanization; the spread of mass democracy; the globalization and international spread of political and economic forms that transformed the structures of government, capital, the state, and citizenship; and the expansion of a public domain that has been so rapid as to challenge the necessity of a private life. Not only have these structural features been important in mapping out the transformation of the modern world, they have also functioned to undermine the modernist ideals of progress, order, and civilization. The mythology of the modern—that stability and order are immanent—is belied by the massive shifts in human practices that show how much change and upheaval are continually taking place.

The ideology of the modern, with its ambitions for order, truth, and universalism, has overshadowed many dissenting, argumentative, and contradictory voices. As a result, we are unacquainted with the antinomies integral to our own actions, what Theodor Adorno called "underground history" and Michel Foucault termed "the dark side" of human history. An important defining characteristic of the modern has been the intellectual separation of the real from the ideal, the experienced from the imagined, the mundane from the theoretical, the actual from the possible. It is this artificial divide that has allowed the contested, disjunctive, and arbitrary events of the everyday world to be minimized in importance, while a bolstering and escalating of the imaginative search for a technically perfectible, orderly world has continued. Consequently, we moderns have arrived at a moment in history when inaccurate ideas—such as the mechanistic orderliness of society—continue to receive tacit acceptance as useful assumptions upon which everyday life can proceed.

The desire for order, it seems, is stronger than our willingness to recognize the disorderliness inherent in the complexity of human affairs. This desire is reflected in the modern habits of codifying knowledge, bordering spheres of thought with disciplinary divisions, and creating elite cultures and hierarchies of value (Lyotard 1984). It is also seen in the widespread use of material goods as icons of status. Ranking commodities into categories, distinguishing them as mainstream or from the ghetto, expensive or vulgar, fashionable or avant-garde, imposes on the world a sense of order. At the same time, these categorizations are attempts to organize the lived experience of human beings in ways that are artificial and full of anomalies. It is this desire for order that stands against the realization that ours is an era of much greater diversity and confusion than is generally acknowledged. The desire for orderliness has produced a concentration on the surface life, on the appearance of things, that has had the effect of concealing other more disturbing features of the modern.

The Surface Life

Georg Simmel, the early-twentieth-century social theorist, concentrated his analysis of modernity on the mundane experiences of urban life. He recognized the ambiguities and contradictions of the city and predicted that the modern individual would fail to develop a rich character in such an environment. The late nineteenth century and the early twentieth century were dominated by the urban phenomenon that grew from an explosion in the population size, migration to the cities to escape a faltering rural economy, restructuring of the built environment, and new work practices imposed on individuals by the timetables of mass production and the mechanization of factories. Simmel (1911, p. 50) understood that the modern epoch hinged on a belief in the legitimacy of the social. That is, the modern individual needed to understand that the social was more important than his or her unique circumstances: The modern individual was to be submerged into the broader society.

This manner of thinking was different from that of the ancien régime in which the individual's sense of him- or herself was as a member of a group, a stable category such as clan, fiefdom, or class aristocrat, warrior, cleric, or peasant. With the breakdown of these traditional social locations, the modern individual needed to look for other anchorages; material wealth was one such source and, in particular, the conspicuous consumption of status items such as fashionable goods. Material consumption supplied a shared language of identifiable goods and values that allowed individuals to see order and commonalities between themselves. Subsequently, when individuals encountered one another in the city—as they did by chance, and without the regulation of class or clan membership—they acted on the basis of putative common interests provided by the display of material goods and a shared desire for orderliness.

An essay by Simmel called *The Metropolis and Mental Life* was published in 1911, well before the gargantuan cityscape had been built and before the Manhattan skyline, with its improbable architectural towers, had become the emblem of the modern. As individuals migrated to the city, Simmel's interest centered on how individuals could endure the actualities of the city, how they could make sense of its size, complexities, and the cacophony it generated. His concern focused on the tension between the actual experience and the imagined, more desirable one.

For Simmel, the essence of modernity was social exchange, which was carried out through "webs of affiliation" and the haphazard intersection of social circles. He described society as a labyrinth, experienced by the individual as a form of ceaseless everyday interaction. There was little about city life that was orderly or psychologically reassuring. Indeed, for Simmel, the city seemed unstructured and unsystematic; it was more like an entanglement. Thus, the new view of the social, which was developing alongside that of the city, was a function of the individual's changed consciousness. The social was now formed from exchange and encounter— it was a contingency, not an absolute. The modern individual needed to develop the ability to think of him- or herself self-consciously and imaginatively as others might see him or her. Living in a public domain meant there was an unending parade of new impressions to be mediated and new demands to be answered, including the demands of a daily routine regimented by economic calculation, the challenges of diverse interpersonal exchanges, and the need to express tolerance of the strangers encountered at close physical proximity. The city required of the individual a rigorous self-discipline and willingness to conform, not to the traditional forms found in the ancien régime but to the complex and ever-changing timetables of commerce and industry.

The diversity and rapidity of the social exchanges taking place generated a sense of constant bombardment, causing the individual to become nervous and skittish. These tensions and frictions, however, were not easily identified by individuals themselves, although their effects were strong enough to make them seek out alternative sources of orderliness and certainty. In an otherwise tempestuous and fickle environment the means for securing this sense of certainty was through material goods, so possessions such as fashionable clothing, high-status commodities, and symbols of power took on new meanings. The pursuit of conspicuous consumption, then, not only functioned as a source of pleasure but also provided a common language that allowed strangers to feel as if they understood one another and lived in the same world. On the basis of these shared material interests, strangers experienced a sense of commonality, though, in actuality, the strains of disorderliness, fragmentation, and contestation were stronger.

This culture of materialism made the modern city seem as if it were a bourse, a place of exchange where commodities of unlike character could be valued on a common scale. Simmel noted that modern individuals living in the city developed an external reserve toward others, a blasé attitude, a slight sense of aversion that sometimes approximated repulsion, and this attitude further enabled interaction between strangers to proceed. As such an attitude emphasized the necessity of regarding individuals as if they themselves were objects, the unanticipated social encounters of the everyday and the complexities of city life came to seem more manageable.

When human sociality becomes dominated by a materialistic ethic, social exchange approximates other modes of exchange where money is involved. As money provides a lingua franca that generates activity and exchange where there may not previously have been any, the members of a population with shared interests in material commodities can be induced to associate with one another more than ever before. When the materialistic ethos of society takes on overriding importance, relations between people and objects alike become readily quantified, categorized, and evaluated. Such activity takes place at the surface of life, employing patterns of conduct and well-practiced routines that make everything seem obvious, simple, and orderly.

Social exchange of this kind gives a false importance to the visible world of goods and commodities while at the same time emphasizing the importance of the surface life. When material goods and consumer items such as cars, foodstuffs, clothing, entertainment, music, and sports are used as a common language or sign system, they create a sense of commonality that exists only at the surface of life. The result is a manner of living in which material goods are used as the building blocks of the social. In such a world, material exchange becomes an overriding interest, and the urgency and excitement that accompany the constant activity of trade create the cacophony that prevents authentic social engagement. Simmel saw the modern individual, with his or her blasé attitude and lack of engagement, as suffering the consequences of modernity, namely, an incapacity to determine values and respond effectively to much that was new and demanding. Life in the modern city was lived by formula, by the norms and imperatives of business and public roles. In turn, this produced an aesthetic or judgmental exhaustion in the individual which led to a confusion and incapacity to discern meaning and value.

The metaphor of society as a bourse enables us to see how thoroughly we have incorporated into social life the materialistic ethic with its privileging of the visible and quantifiable. Mass industrialization and the urban experience have provided the backdrop against which a sociality based on the obvious has developed. By living at the surface, by engaging with people and ideas as if they were material commodities, it has seemed as if the goal of orderliness could be realized. The individual's growing expertise with handling the material, his or her increasing confidence in manipulating goods to meet his or her own ends has the effect of making the obvious, the material, the quantifiable, seem highly attractive. This emphasis on the material distracts the modern individual from recognizing other aspects of the social, namely, the conflicts, ambiguities, and mysteries that also constitute the human experience. Such attempts to simplify the social experience by imposing a materialistic system of order have the unhappy consequence of increasing the individual's inability to deal with ambiguities and the problems of disorder.

The growth of the city has been a great influence on the development of these shallow sensibilities. The city has provided the environment in which the new culture of conspicuous consumption could expand to include all levels of society. It is the cornerstone of new industries designed to manufacture, market, and advertise what has become a culture of materialism. These new industries, which are concerned with the mass media, Hollywood, entertainment, fashion, tourism, and advertising, induce in the individual habits of rampant consumerism by promising an ever more orderly, secure, satisfying, and materially comfortable life (Horkheimer and Adorno, 1973). As the materialistic ethic gains momentum, we are brought

abruptly to the point of being unable to differentiate among objects, qualities, and values. As Simmel foresaw, and Umberto Eco (1986) and Jean
Baudrillard (1988) have more recently articulated, we have no idea what is
real, what is fake, what is valuable, what is a reproduction. There are so
many images of reality that reality itself becomes a simulacrum, an imitation. We live in a world blanketed by hallucinations, a simulated world in
which television characters, rock and roll musicians, and movie stars are
more appealing than the real people in our neighborhood. Although these
images and icons provide a language for daily social exchange that we
find useful, they also lead us further away from engaging with the real
nature of the modern social world. The simulated character of a social life
lived at the surface ignores, or at least dilutes, the existing undercurrents
of friction, contestation, and difference. In short, this emphasis on materialism, of evaluating objects, people, and qualities as if they were commodities traded in a bourse, has concealed the unanticipated consequences of
living at the surface, namely, the moral vacuity of a consumerist ethic.

The Desire for An Orderly World

In the twentieth century, the unprecedented reach of Hollywood, coupled
with the bizarre logic of the Gulags, international trade and banking,
colonialism and nuclear proliferation, suggests that the world is riven by
greater ambiguities and contradictions than the ideology of the modern
has overtly indicated. To nullify these deep conflicts, individuals are drawn
to the surface of social life, to the level of display and spectacle where all is
thought obvious and unequivocal and where there are few intimations of
any depth. This is a world of commodified goods and ideas, a world of the
despotic banal where even the physical body and its appearance have
become servants to the sense of orderliness and certainty sought after in
the obvious.

To participate in the bourse of social life, to enjoy easy and productive
exchange with strangers, individuals must appear and act in expected ways.
We must tailor ourselves to fit the others' desires. In describing the types
of social characters we have become, such as the dude or the fashion plate,
the adventurer or the spendthrift, Simmel (1971) illustrates how we have
commodified ourselves and made ourselves into standardized types that
represent expected qualities. Our willingness to groom ourselves in accord
with continually changing fashions and customs demonstrates the
preeminence we give to the obvious, to the visible details of appearance.

The importance of appearance reflects the modern privileging of the visible over that of the inferred and abstract. This attitude has drawn our attention yet again to the appearance of things, to the surface life, where the specificities of image and impression predominate. The irony of the times is represented to us daily, through the jungle of images and commodities that constitute the rich material culture in which we live. As we view another advertisement, watch another film, attend another spectacular event, we are further embedded in the artificial reality of the fabricated world, the hyperreality described by Eco and Baudrillard. We come to know ourselves and one another as if we were characters in this hyperreality, and, subsequently, our knowledge is circumscribed by interests other than our own. As we accept the products of the culture industries, we are slowly fused into a populace that shares a passion for a passing parade of hallucinated Hollywood situations. The success of the image heralds the destruction of the social because, in effect, it has made individuals into an audience witnessing the spectacle of society but not participating directly in its manufacture. As a result we have been severed from the workings of society.

Thus, in the late twentieth century, for all our sophisticated knowledge and technical virtuosity, we moderns are too much enthralled by the appearance of things. The reason we are so underlines the retreat to the surface of life that we have accepted. The past is represented as the essence of an orderly future. The long-lived desire for a simple, orderly world— that hallmark of the rational, purposive, modern society—can be seen as the impetus behind our refusal to recognize the importance of friction and the disturbing acts of irrationality and antinomy that occur daily in the modern world. Our wish to inhabit a rational world has encouraged the growth of a materialistic ethic that, in turn, supports a misguided belief in the possibility of a more orderly future. In effect, the desire for order and greater coherency has concealed much of the evidence to the contrary. Now, it seems, the common experience that seems to hold us together is a desire for the obvious and simple, and this desire produces a kind of moral and intellectual vertigo that prevents us from seeing the chronic conflict and confusion that exist in the modern world.

References

Baudrillard, Jean. *Selected Writings*. Stanford University Press, 1988.
Eco, Umberto. *Faith in Fakes*. Secker & Warburg, 1986.

Foucault, Michel. *Discipline and Punish*. Allen Lane, 1977.
Horkheimer, Max & Adorno. Theodor *Dialectic of Enlightenment*. Allen Lane, 1973.
Lyotard, Jean. *The Postmodern Condition*. University of Minnesota Press, 1984.
Simmel, Georg. *The Sociology of Georg Simmel*. Free Press, 1950.
————. *On Individuality and Social Forms*. University of Chicago Press, 1971.

12

Stay in My House

Kaj Nyman

I

During the nineteenth century, industrial production replaced agriculture as the major source of income in Western Europe and North America. Large transfers of population became necessary. The manner in which buildings were constructed thus acquired a new social importance: The provision of housing became a production factor.

Architects were late in realizing what this meant for architecture; engineers understood the matter much better.[1] Classical architecture stood in the way of the development of society. There was a risk that the initiative, with regard to the direction in which architecture should develop, could slide over from the architects to the engineers.

Functionalism may be seen as a rational solution to this problem. Instead of becoming engineers and losing their distinctive character, in functionalism the architects found a way of surpassing the engineers in social orientation. Once all buildings were defined as having social importance as well as being architecture, it became possible to find a common denominator: The buildings were adapted to the demands of society when designed according to the new functionalist doctrines. Architecture was immediately transformed; what had been an impediment to cultural development was now the lubricant that greased the machinery of society.[2] This is where architecture stands today, with a few exceptions.

What does it mean for architecture that social processes are given their form by the predominant culture, such that the human justification of prevailing attitudes is not questioned? After the breakthrough of functionalism, architecture has played a role quite different from that of an art form that makes everyday life into poetry for each and every person who is open to stimulation. Its purpose now is to maintain by force the social signs that everyone must obey if the preferences of the culture are to be carried through. Functionalism completes its mission by reducing human life to unambiguous functions defined by society, after which it tries to realize the equivalent between these and the material form of buildings. Architecture becomes a type of mass communication that clears away people's prejudices against that which is culturally necessary and restores their faith in the "collective knowing" of what is best for the individual. Manifoldness in the human experience of living and building is replaced by an unambiguity that is a prerequisite of control over the social processes. In order to avoid risking the survival of his culture, the individual is forced to refrain from the intentions deep in his soul and instead embrace those of society.[3]

For the architects, the successful change of direction meant that they achieved a position of power that they had not previously been able to dream of. Given the division of labor in society, architects were charged with the planning of all buildings; that is, they became masters of that area into which the major part of the surplus from the economy was channeled. A central position, indeed! This status can be maintained only through continual concessions. The architects are subject to unceasing control to prevent their artistic will from taking over again. An art of building that was freely available to people in their everyday lives would be as much a danger to the survival of the existing cultural values as a situation in which every individual could freely decide about his medical care or his child's basic schooling or make agreements with his fellow man without the aid of lawyers. The price for the architect's acquisition of such power is that his profession degenerates to the work of a technician.

II

The principle of the existing cultural order is that of competition; in other words, we must be permanently prepared to choose the correct side and thus acquire the right to exploit the other side. When competing, speed is invaluable—it is a matter of restlessly producing counter measures before our opponent has had time to attack. It is in this game that the architect is expected to make his contribution.

In nature there is no competition (in the human sense of the word), only adaptation, and thus there is no "speed"; everything happens "at the right time." The principle here is that the organism humble enough to learn what it means to be a part of an environment is the one to survive.[4]

"Human nature,"reflected in the psyche typical of the human species, has precisely this function. The many thousands of years of experience that are stored there, help us to adapt to changes in our environment in a way that fosters the survival of the species.[5] Part of this experience concerns houses and building. It is largely shared, at least by those people who belong to the same culture. The experience has an authority that prevents the architectonic elements from flowing out in ever new directions, risking the environmental adaptation that has already been achieved. Is this psychic mechanism still working? The dominant cultural order sometimes seems to be doing its best to battle against the very existence of such "inertia."

Natural systems are characterized by their autonomy. The traditional architecture that we admire all over the world reaches its artistic form without any conscious control. Life—both physical and mental—knows its boundaries and has an inherent ability to maintain its form.

Evolution takes place in open systems as a result of changes in the environment. In nature, this adaptation almost always happens in a way that does not demolish the form previously reached. Our Western culture is different: It has lost control of its way of evolving, thereby causing deeper and deeper changes in its own environment, which presupposes an ever quickening evolution, and so on.

All evolution takes place at the cost of a decrease in autonomy. A development based upon speed has serious consequences in that the breaking down of autonomy cannot be compensated by the specialized organization (increasing "intelligence") that evolves; rather, the natural hierarchy of mutual dependence must be replaced by a power hierarchy. The loss of autonomy implies "friction," which requires "lubrication" through increased supervision. Instead of increasing manifoldness, a far too quick evolution causes qualitative information to be lost.[6]

Public planning implies distrust of people's autonomy—a sort of negation of human life. (This is not to say that such planning is dispensable, however.) In the past, inertia was accepted as the necessary precondition of autonomy. Houses were what they always had been: a limitation of possible forms of behavior but also, at the same time, a source of stability in life, providing security. Modern society wants to be rid of all limitations upon its freedom of movement and does not realize the usefulness of its context.[7] Instead of mustering their strength in the "struggle for existence" that yields pleasurable knowledge about the conditions of existence, people have used their intelligence to free themselves from contextual restriction and now waste their strength on mutual competition.[8]

Among the "skilled products" that engineers are busy developing is the "intelligent building"—a constructed environment that liberates its inhabitants from every exertion, whether physical or mental. Yet again, the initiative is in the process of slipping away from the architects and moving toward the engineers. Simultaneously, the architect's role of mercenary responsible for public order, is emphasized. By accepting the computer-driven concept of the "intelligent building" as a paradigm, the architects effectively prevent people's opinions from becoming too fragmented from a cultural viewpoint.

As economic integration progresses and the totalitarian features of culture become more predominant, it is increasingly important that the architect represents the "correct" attitudes directly, on the basis of his profession. To question the justification of the prevailing attitudes is no longer compatible with the architect's role.

III

What we are, as people, has been learned over perhaps hundreds of thousands of years through a continuing reappraisal of earlier experiences in a permanent self-reference. Very gradually, irreversible psychic structures have been formed that we are generally no longer aware of; we call them habits or customs. Only that which after many trials is seen to have a lasting capability to survive is constituted as habit.[9]

The part of our behavior that is based on routine and habit—the presence of which is a precondition for what is known as attention—has its stronghold in buildings and towns. That which is specifically human can thus be said to have become formed within them. The architectonic form is an expression of human form, or, put another way: Through architecture, human characteristics are given material expression.

Inasmuch as architecture is a realization of the strength of inertia in human habits, it implies a retention of what is specifically human. Such a function is ideally suited to architecture, which came into being through a long, unsurveyable, and inflexible process. The will of modern architecture to modernize society implies that it denies the importance of architecture as a manifestation of human form.

Form exists only as language.[10] It was not until desire was stabilized in language that man could express himself and communicate his conceptions to others. Before that, man had no form.

Architecture is human form inasmuch as it is a language for those who use it. An architectural structure that subordinates itself to the demands of culture in terms of unambiguity and controllability becomes for its users a language distorted by incomprehensibility and loses its human form.[11]

That architecture is human form can now be explained in the following way. A major purpose of every language is to retain, by virtue of its inertia, those meanings that have been shown to be vital during the long learning process of the human species. Because we can express only that which has taken form in language, language itself is characterized by a basic inaccessibility: There is no place outside language from which it can be controlled. When we use a language, we are dominated by it. Languages thus protect that which is human.[12] Only he who subordinates himself to a language can occasionally gain insight into the human secret.

Architecture is a language for communication of human meanings that cannot be conveyed in any other manner. We rely upon this language as long as we can see its connection to our lives. It is a language for expressing relations between a person and a built environment and among people in the environment.

The continued existence of environmental relations is vital for every living species—including man, even though he has learned to use his intelligence to overcome problems in environments that are not suitable for him. In fact, environmental relations are a central part of man's humanity.[13] An architecture that has lost its humanity provides inaccurate information about man's environmental relations. It becomes an obstacle in the way of the reproduction of those environmental relations.

IV

Form is not the same as unchangeableness. What persists when form changes could be called the pattern. The change, the arrival of the new out of nothing and without the form disintegrating, can take place because identity and difference are not opposites of each other. Change involves recognizing that which is different but, simultaneously, new identity comes into being. In fact, with respect to open systems (and almost all systems are open), the form can be seen to be holding its ground by changing.

The typical example is capitalist culture. The paradoxical reason for its destructive method of evolving is the necessity for the culture to remain the same: The ruthless dynamics in the development of society are the precondition for the continued existence of Western culture. That which is really new, that which questions prevalent attitudes, is seen by the culture as a threat—because difference comes into being, and difference is friction in the machinery of society. Difference is opposed by every means: The method of science is reduction of the manifoldness of reality, art is enclosed in museums and concert halls or purchased and transformed into possessions, and architecture must follow the "style" of the times.

Creating something new is always problematic: Every creative act is a potential threat to the existing order (as well as a promise of a new order); where no renewal takes place, however, everything loses its meaning over time. What is sufficiently valuable for it to be allowed to become an obstacle for any prospective new breakthrough? How can one know if something new is sufficiently valuable as to be worth risking the order that has already been achieved, perhaps after great sacrifices?

There would seem to be a connection between hanging on to what one has and devoting oneself to something new: between inertia and the unexpected. This connection holds true not in Newton's conception of the world, where room for innovation is lacking because everything, in principle, is predictable, but in the worldview that is developing from the basis of what (somewhat misleadingly) is called chaos research. The new physics sees, in the inertia of nature, a source of unexpected events, of random deviation from what has hitherto existed, which can be a beginning of something qualitatively new.[14] The Creation is still continuing!

Culture is no different.[15] The inertia found in language is the very source of creative actions that now and then light up and reveal to us new possible ways. It is an old truth that the strict rules of languages are the prerequisite for linguistic innovations; the rules themselves, in the collision with a reality that does not allow itself to be tamed, seem to give rise to the deviation from the rules.[16]

One implication is that the ancient, human qualities that have their form in languages are in some way the cause of every new form in which human qualities appear. But it also means that a culture that has lost its faith in language, and wishes to replace the human form with its own shortsighted rationality, cannot create anything new and therefore has no future. Or, expressed somewhat more clearly: The capitalist culture produces something new only within the confines of its own narrow rationality, and it experiences the richness in the human languages as something irrational. Culture, which is the form of society, thereby alienates itself from the people in society. In the long run, this is an impossible situation.

Modern architecture seems to be caught in this trap: To the same degree that architecture subordinates itself to the prevalent culture and becomes an expression of the culture's rationality, it loses its ability to express that which is human. Such an architecture sooner or later becomes too absurd to be able to exist at all.

V

Continuity in every living order demands distrust of changes, but the order loses its meaning if it becomes an end in itself and stiffens into rigidity. With our human limitations, we see this as a problem; for nature it seems to be a principle. When it created man, nature used a basic idea that amounts to precisely this: language. Every language is characterized simultaneously by the inflexibility of its rules and by permanent renewal.

Our culture believes that it can reign supreme over the languages. Hence it opposes its creator and directly calls into question the fundamental idea in Project Man. That is why the architecture that came into existence after the breakthrough of functionalism is often experienced as nonhuman: Instead of expressing what sort of people its users are, that architecture has become a language for what the culture wants—the supremacy of technology, the authority of money.

Is there another possibility? It is apparent that architecture exists on the culture's terms. Its language is limited to the world of nonhuman culture; therefore, it is a foreign language. The precondition for another architecture is that the culture becomes another. Unthinkable?

Not at all. That we can experience architecture as nonhuman is a clear sign that the "other" culture exists. It has always existed: It appears every time we identify ourselves with life, with our nature as a living species, instead of with a culture that is in opposition to nature. Identifying oneself with life does not mean returning to nature. On the contrary, it is using the freedom that is the very essence of what it means to be a cultured being.[17] Human existence means permanently choosing between prevailing cultural attitudes and extremely old human experience.

For the creator of architecture it is a case of seeing through the cultural necessities that prevent human nature from showing itself in the architectonic manifestation. The architect's practical problem, then, is that his imagined creation must be given material form in the cultural reality. (This is not a little problem; certain compromises will be necessary at every turn).

Jean-Paul Sartre had a concept of the human world that is relevant from the standpoint of architecture. His term *pratico-inerte* refers to the inertia that man, as long as he has existed, has unremittingly generated through objectifying concepts and ways of thinking, habits and rituals, tools and machines, towns and buildings. It also refers to the fact that every separate objectification is a manifestation of man's freedom. Hence *pratico-inerte* could be a reflection of this freedom.[18] The creator of architecture would thus be an instrument that is played by mankind's experience. The rules for playing are extremely old; what the music will be like, nobody can foresee.

[Translated from the Swedish by Rodney Bradbury.]

13

Unpredictability, Frictions, and Order

Åke E. Andersson

A Chaotic House

Houses are normally safe, stable, and predictable. In an earthquake the falling down of houses is therefore judged to be a catastrophe. But the stability of a house is physically nothing to be taken for granted, at least not for the theoretically inclined.

We can easily imagine how it would feel to be living in a house of strong and immediate interdependencies between the inhabitants, on the one hand, and the walls, roofs, doors, and windows of the house, on the other. Let us assume that we are visiting such an odd house, equipped with many entrance doors. Upon entering it we suddenly realize that our first step into the house slightly displaces one of the windows of the room (and maybe the position of a ceiling or floor in some other room). Not having anticipated this strange behavior of the house, we would feel uncomfortable and maybe slightly curious. But by moving around the house and carefully observing the consequences of each individual move, we could discover all the interdependencies between inhabitant and house. Predictability and a feeling of safety would be feasible in such a situation of limited interdependence.

Now, let us assume that two persons are entering the house at the same time through two different doors. The first person might change the position of the floor for the second person, whereas the second person might change the position of the roof for the first person. Not knowing about each other, they would have extreme problems detecting causalties. If at some stage they meet up with each other and proceed to communicate, they could by trust, communication, and experimentation successively move toward a full understanding of this new and much more complicated interdependency problem. Predictability could be achieved in this case as well, but the efforts involved, compared to our first example, would be much greater.

Increasing the number of entrants into the house drastically reduces the practical and theoretical possibilities of detecting the deterministic interdependencies between the people, affecting the geometrical structure of the house. Increasing interdependencies (nonlinearities) thus necessarily increase the problems of prediction by increasing the complexity of the dynamic process. With many inhabitants, the house would be inherently chaotic. It would be deterministic, bounded, and yet unpredictable. Chaos and unpredictability do not breed feelings of safety and reliability.

Fortunately, normal houses do not behave in this way. Even children, who frequently slam doors and run through rooms, do not normally cause displacement of roofs, windows, or the doors themselves. The physical depreciation of the house is sliding on a qualitatively different time scale, quite undisturbed by such fast irregularities of the inhabitants.

This model (or maybe metaphor) can be generalized into a theory of dynamic processes on different time scales.

Speeds of Change and Scales of Time

Differences in speeds of change are inherent in the mixing of different physical processes. Geological change occurs on a time scale that evades all possibility of our direct observation. Within his lifetime, a ski instructor living in Zermatt can wake up every morning to discover without any surprise that he can still see the Matterhorn in the same position of an unchanged alpine landscape, despite having read an article claiming that the Alps belong to the most dynamic parts of the global geology. Similarly, a bug, living on the skull of this Zermatter can comfortably spend the days and nights in the hair forest, because the process of getting bald will proceed at a much slower pace than the life span of an ordinary bug. And on the body of the bug there would surely be some bacteria and other microorganisms that would behave as if that habitat were dynamically stabilized. What counts in all these examples is the difference in speeds of change, *relative* to each other. A fast process can be predictable only if it occurs within bounds, given by some slow process. Differences in time-scales, then, are essential to the understanding and predictability of inherently complex dynamic systems.

Collectivity and Order

But such differences are not enough to give order and predictability. At least some of the slow variables must have a public or collective character, if ordered and predictable patterns are to occur. Any observable dynamic system tends to have some order-creating variable with collective effects. In a physical system the average temperature of a geographical region might give a high degree of predictability to some, otherwise completely unpredictable biological phenomenon. In the northern Swedish city of Umeå, for example, around May 22, all birches instantaneously and in a seemingly highly coordinated way, go from gray to green within a few hours. In all of Sweden, more than 80 percent of the adult population participate in the general election on the third Sunday of September, every four years. And every morning, in every metropolitan city around the world, people are agglomerated in the same transportation mess for approximately the same duration of time, as if ordered to those highly congested stretches by some omnipotent dictator.

In all three cases, there is no omnipotent decisionmaker forcing uniformity upon the micro-units. Rather, social and economic frictions are giving rise to slow and sufficiently collective variables, characterizing the dynamics of the social system.

Understanding Social Dynamics

R. Thom (1975), I. Prigogine (1977), and H. Haken (1977) are three of the main scientists who have used these theoretical concepts to improve our understanding of the physical, chemical, and biological processes that formerly were seen as examples of nondeterministic (i.e., random) processes subjected to basically unpredictable shocks. Their contributions have made it possible to predict the formerly unpredictable, to understand what was hidden behind apparent randomness, and to distinguish between structurally stable phenomena and those that are unstable in terms of form and organization. Furthermore, we have opened our minds to the importance of bifurcations in biological and ecological processes and have given explicit content to seemingly hollow concepts. One example might suffice. It is often said about certain periods of history that "the time was ripe," or that such periods were great or even evolution altering. Haken, Prigogine, and Thom have given explicit and very precise conditions for when "the time is ripe."

It is of course possible that concepts and theories developed to give deeper understanding of biological, chemical, and physical stability and that changes are not at all relevant for the understanding of social systems. I am not of that opinion. Rather, I believe that differential time scales and the publicness of certain phenomena are essential for the understanding and predictability of complex social systems and their evolution. Unlimited dynamics in a frictionless world is impossible in the social structures that have emerged, that we daily observe, and in which we live our lives in a mostly predictable way.

Any variable of a social system can be classified in one of two fundamental dimensions. One of these dimensions has to do with the scope of consequences for other variables. A brief example can clarify the issue. A hamburger at a McDonald's outlet is a highly private variable (unless the consumer happens to be Madonna). The recipe, including the logistical principles, does not have this character of privacy. Rather, it has an enormous collective or public scope and is sent around the world to be used in every McDonald's outlet and among all the imitators of McDonald's, including a large number of private households. Similarly, there are recipes for making cars, pharmaceuticals, and all other consumer and most producer goods and services. There are even recipes for making recipes. Goods are private and the recipes behind them are public. Or, put somewhat more generally, artifacts are private but the knowledge is public. This principle holds in spite of the attempts at scrambling knowledge to achieve property rights for some limited period of time.

Similarly, there is a fundamental difference between behavior and rules of behavior. Behavior is mostly private, whereas rules of behavior are mostly of a public nature. When the Parliament takes a decision on a new law, it is assumed to regulate the behavior of those concerned within the population of the country.

There are also time-scale differences between variables. Driving the car to work in the morning involves actions on a fast time scale. Taking decisions on traffic laws and building roads and streets for traffic involve decisionmaking and actions on an enormously slower time scale.

Rules of behavior, knowledge, and networks are variables that are of extreme qualitative importance because they are inherently public *and changing on a slow time scale*. Together they constitute the most important variables, forming the common *arena*, constraining the *games* to be played by millions of individual actors in their pursuit of individual goals, targets, or whims of the mind. It is a meta-stability and the ordering capacity of the arena that contribute to the predictability and planning of the private and fast games of life.

Creative Destruction

The fact that certain processes are very slow and have a capacity to order behavior into predictable patterns should not be confused with the idea that these conditions are stationary and that causes and effects are unidirectional. The games will in imperceptible ways have an impact or backfire upon the arena. There will always be destruction, and this destruction tends to be creative, in the sense that it shapes the conditions on the emerging new arena. The history of the Soviet Union provides an example. At the end of the 1920s the Soviet Union engaged itself in a massive industrialization program. Based on theoretical models by a group of young economists, working on Marxian foundations, an industrialization model was built, stressing industrialization in stages and investments initially in heavy industries. Thereafter, the investment program would shift toward lighter and lighter industry, agriculture and services. Marxian economic theory did not contain any capital accounting principles. The only things that matter in this theory are investment and production, i.e., one-period variables of a basically individual character. Technology, infrastructure, and the slow and steady upkeep of capital to ensure long-term sustainability of the economy and ecology did not belong to this set of ideas. Slowly but steadily this lead to a hollowing of the Soviet economic system. The public capital, and especially all the networks, deteriorated until it was a large but extremely brittle structure that could not be relied upon as a stable arena for daily production and exchange. The system had to disintegrate into an anarchic set of basically isolated and rather primitive economies. Mathematically, it can be proved that a slow and steady deterioration of public variables will sooner or later entail a dramatic bifurcation from one equilibrium situation (or stage) into a structurally different equilibrium situation. The creative destruction is a slow evolution inevitably leading to revolutionary change, once the limits of the arena have been reached.

A bifurcation of this kind need not be negative in terms of welfare development. The industrial societies in advanced economies like the United States, Japan and Western Europe have been involved in a creative destruction since the beginning of this century, thus preparing for a rapid structural change into postindustrial societies. During the transition from the nineteenth to the twentieth centuries, most of the currently industrialized countries were in the process of completing their stage of transformation into industrial societies. The basic characteristics of these late industrial societies were exploitation of division of labor, use of mineral energy sources in different kinds of machinery, use of railroads for intracontinental transportation, and use of machine-driven ships for intercontinental conveyance of goods. Trade in commodities acted as a substitute for transportation of labor and human capital.

The transportation system—or, rather, the logistical network—was extremely sparse and based on the exploitation of economies of scale. In fact, all of the networks determined a metastable arena, forcing industrial and regional organization of production and trade into a rather determinate, efficient pattern. The characteristics of this arena required mass production of relatively uniform products in large manufacturing towns, located at the point of tangency between train transportation and shipping networks. Detroit, with its production of the standardized model T Ford, exemplified efficient market adaption to the industrial arena.

It was obvious to the industrialists of this early stage that many human and natural resources could not be exploited if the production system had to be adapted to such a sparse set of networks for transportation of goods and communication of ideas and information. As early as the first few decades of this century, telecommunication and road networks started to fill the empty spaces in the web of railroads, canals, and rivers with roads, highways, and streets. By the 1930s, the networks for car and truck transportation in most countries were much larger than the older networks. By the 1960s, with the completion of the American highway system, it was obvious that the industrial society had, through the creation of new networks, destroyed the arena on which the industrial organization had been developed. There was a serious mismatch between the arena and the organizational game that now dominated the market games. New corporate structures and new locational patterns were emerging that had little in common with the classic industrial corporation with its gigantic plants located close to the waters and the main railroads. Instead, corporations like IKEA, Digital Equipment, Apple, and Benetton were growing in dispersed mushroom patterns with no obvious headquarters and with a product structure based on *scope* rather than *scale*. We were moving into a postindustrial society or, rather, a C-society (i.e., a society based on exploitation of the new Cognitive, Creative, and Communication capacities) that had superseded the resource exploitation possibilities associated with the sparse industrial arena. The slow and steady improvement of the early industrial arena had led to the destruction of the topological characteristics of that arena, and this destruction inevitably became incompatible with the industrial organization that had initiated and financed the new level of knowledge and new structure of the networks.

Evolution by Constraints versus Revolution by Bifurcation

Evolution is by definition a smooth and thus essentially predictable trajectory. By small and almost continuous additions and amendments, the system under study transforms itself so that each consecutive snapshot representation of outcomes is similar to any preceding or following snapshot. It is like a sliding equilibrium process. Evolution defined in this way requires a constraining arena prohibiting large disruptions by inertia.

Revolutionary change is not ruled out in my theory of dynamics. Although it is possible to prove that arena game models exhibit evolutionary equilibria most of the time, they also contain infrequent but absolute necessary bifurcations of a revolutionary nature. Evolution implies revolution. But the infrequency of revolutions is the saving grace of the theory.

It might seem from the preceding sections that my view of evolution as caused by the subdivision of a dynamic system into arena- and game-variables implies a strong belief in predictability. It may also seem that we would minimize, at any cost, the number of bifurcations or the relative risk of bifurcations within an economy or a social system. Admittedly, there are advantages to living in a system that is easily observable, predictable, and, at least partially, controllable. In a world without these properties, private and public policymaking would deteriorate into operational and tactical reactions to unexpected accidents and windfall gains. But it would be equally problematic to envisage a world where everything is constrained by inertia and other dynamic factors, resulting in a stagnant society where everything could be predicted and no action would make any difference. Somehow there is a need for an in-between state where dramatic changes can be pondered. And the zones of bifurcation are those short instances in the evolutionary process where nothing is predictable and everything is possible. The unpredictable zones of bifurcation provide the opportunities of dramatic change but also the fundamental uncertainty of the future.

The times of bifurcation are those during which the otherwise stabilizing and well-structured arena unfolds into a flat land where the evolutionary flows can take on any slow routing and where even a seemingly strong and steady unidirectional flow can make drastic turns as a result of even very small changes of some parameter. When time is ripe for a bifurcation, small actions matter. In times of bifurcation, people interested in action can get remarkable responses at the expense of little energy. And with skills much can be achieved under these situations of structural instability and lack of inertial constraints.

The situation I am portraying is in many ways similar to the situation facing a skilled pool player. His ideal arena is a billiard table without any structural features other than absolute flatness. In order to control the situation, the player needs the unlimited possibilities of bifurcations, whereby every little variation in angle, direction, and force can have a major impact on all the trajectories of other balls on the table.

Periods of artistic, scientific, and political creativity are times of structural instability with a large potential for bifurcations. Paradoxically, they are also times of extreme dependence upon intellectual frictions. It is during such periods that creators find strong reasons for returning to almost forgotten artistic ideas, scientific theories, and philosophical paradigms, preserved only by frictions imbedded in the construction of the intellectual world with its libraries, museums, and invisible global networking of scholars. History provides convincing evidence that new ideas tend to develop and spread rapidly when a pronounced shift occurs from one steady evolutionary stage into another—that is, when the society at large is undergoing a phase transition from one equilibrium situation into another.

History also tells us that such paradigmatic phase transitions are times of Renaissance. Determined creative action requires ambiguous times. We are leaving the highly predictable last stage of the industrial society. We are slowly but steadily moving into a fully developed C-society—a society fully exploiting human- and computer-based cognitive capacity, individual and social creativity, and the new possibilities of contact and communication between people. During the transformation into this new society we will be victims of structural instability and unpredictability, as well as creating new arts, scientific knowledge and new characteristics of the political systems. This we can foresee, and very little else.

Slow and Fast Processes

The subdivision of processes of change is doubtless extremely useful in the analysis of physical phase transitions, bio-chemical transformation e.g. in auto-catalytic processes, and in biological birth and death and other situations of biological transformation. However, I am absolutely convinced that such a subdivision into different timescales to capture differences in degree of resistance to change is of even greater importance in the analysis of the changes that have occurred in the economic history of Europe. One historical process is especially revealing.

Pirenne has claimed that the great transformation of Europe in the late Middle Ages were primarily the consequence of a slow and steady improvement of the basic conditions of trade. In our terminology the very slow, friction-ridden but steady change process of improvement of the transportation infrastructure of Europe paved the way for a major and very dramatic transformation of the possibilities of division of labor between different parts of Europe and a rapid consequential increase of trade and transportation. Division of labor, communication, transportation and trade are in this context different aspects of the same phenomenon. Without division of labor no trade, without trade no possibilities of division of labor. But trade cannot occur without transportation and transportation is not possible without the availability of a network and a sufficiently efficient technology of transportation being available. The dynamically naive would be worrying about the "chicken or egg" of this causal process. No such question can occur if we allow for possibilities of subdivision according to typical speed of change. The slow process of establishing a shipping technology and an associated network must reach some critical stage of completion before any major trade, transportation and division of labor can occur. Thus, the causal chain suggested by Pirenne must be the right one. Indeed, the large-scale growth of new (commercial) towns was *caused* by the slow and steady reduction of frictions on trade and transportation flows.

The French historian Fernand Braudel (1979) has argued that the Renaissance and similar creative explosions have always been secondary to economic expansion. Such a claim has a certain ring of historical materialism to it. However, it is also possible to see this claim in the context of the Renaissance of the fifteenth century within a slightly different explanatory chain of reasoning. The Renaissance was a highly localized phenomenon, occurring primarily in the commercial cities of northern Italy but with a focus in Florence and in a few cities belonging to the Hanseatic network, notably Bruges. It is also clear that this creative eruption was of quite short duration and that it occurred in a period characterized by economic structural instability and disequilibrium.

Income and wealth were growing very rapidly during the process of urbanization at that time. New economic, political, and cultural relations were established among the new commercial urbanites. In fact, enormous wealth was accumulated among the new entrepreneurs; so nobody could claim that this wealth accumulation process was following any balanced path. The first logistical revolution provided the means by which the uneven accumulation of wealth occurred. But this was not a sufficient condition for a creative expansion. Rather, an analysis of the interaction of fast and slow processes in conjunction with the facts of history would suggest that the creative process of the Renaissance, once sufficient wealth had been accumulated, was caused by structural instability that allowed for radically new combinations of ideas in the arts, architecture, politics and science. Such an explanation of the very rapid transformation of Western Europe during the late Middle Ages and the Renaissance suggests a more general explanation of outbursts of creativity in circumstances when *both time and place happened to be ripe.*

Similar patterns have been suggested in explanations of the interaction between the slow and steady change of infrastructural conditions and the occurrence of rapid economic, political and cultural transformation. In McClellands' analysis of the transformation of Greece in the period from 500B.C. to 300B.C. there is an obvious correlation in his data between the establishment of new possibilities of trade, the very rapid and unbalanced growth of trade and other interaction, and the creative outburst occurring in this process of rapid structural change of the economy.

It is as if the phase transition from one stable structure into another providing a period of complete structural instability is the breeding ground for all kinds of creators and their systems disrupting activities. And this is presumably the reason why all major creative processes tend to be of short duration. When the timescales once more depart and the friction-ridden arena once more becomes clearly defined, everything goes back to an equilibrium with structural stability. Then creativity begins fading away. Frictions are important determinants of structural stability and order, and could therefore be seen as an enemy of creativity. But the paradox remains— friction is not only one of the conditions of structural stability and preservation of order. At the same time the frictions of knowledge is a necessary condition for individual and social memories and thus a precondition for meaningful creativity.

References

Braudel, F. *Les Jeux de L'echange*. Paris, 1979.
Haken, H. *Synergetics. An Introduction*. Berlin: Springer Verlag, 1977.
Pirenne, H. *Economic and Social History of Medieval Europe*. London, 1936.
Prigogine, I. *Self-Organization in Non-Equilibrium Systems*. New York, 1977.
Thom, R. *Structural Stability and Morphogenesis*. Reading, Massachusetts, 1975.

14

Friction and Inertia in Industrial Design

John Heskett

"Friction," says Nordal Åkerman, "is what keeps you from realizing your goals. . . . It constitutes the divide between dream and reality" (see Chapter 1). The opening remarks in a paper by the American designer Jay Doblin, which he worked on at the end of his life, echoed that perception somewhat. He stated: "Designers are frequently frustrated by a recurring phenomenon: consumers who should know better often seem to choose products of inferior design."[1] Yet although Doblin sought with great intelligence to analyze this phenomenon, he never explained it. There is at least a prima facie case for investigating the concepts of friction and inertia in the field of industrial design as a potential means of comprehending this gulf between aims and achievement that is so frequently left unexplored.

In theoretical terms, every designer is concerned with the future. That is the nature of the designer's work. At drawing boards, in workshops, and, increasingly, at computers, designers are concerned with ideas that will become the tangible products, communications, environments, and systems we will all use. Whether it entails a pamphlet for production in two weeks' time, a product for a year from now, or a major communication system for three years hence, work done in the present always has an innate orientation to the future.

In this process, there is of course always the possibility of change and, potentially, of improvement. Indeed, throughout the history of modern design, designers have believed and argued that change is not only possible but desirable. Moreover, the changes envisaged have not only pertained to design (e.g., physical artifacts, communications, and environments), but have extended into the social, economic, and cultural spheres as well. In Europe, examples can be found in the work of William Morris, Herman Muthesius, and Walter Gropius; and in the United States, in such different sources as the Shaker movement and the work of Frank Lloyd Wright and more recently, Victor Papanek. All of these designers, in one way or another, have manifested a clear sense of the relationship between designed form and social behavior.

A major problem with realizing the possibilities of these theoretical programs for change through design, however, is evident in the context of design practice. Contrary to the assumption in much design theory that design practitioners are free individuals able to determine their own course, most designers work in some form of organization where the scope for independent decisionmaking can be highly limited. Designers also work under widely varying conditions, and they are subject to very different kinds of control by nondesigners, who have their own perceptions of design and the purposes it should serve.

Designers therefore face considerable uncertainty in their role in terms of what it could and should be able to achieve. In a short chapter it is, of course, impossible to discuss the whole field of design. Accordingly, in the following exploration of the concepts of friction and inertia, I will focus on just this one aspect of designers as change makers, including a discussion of the problems arising from evolutionary changes in the way design has developed.

The Role of Design

Any recognizable area of practice needs to have a common understanding of what it is and does, what it should achieve, the processes by which this can be accomplished, and the values embodied in it. The task is eased if an area of practice becomes so secure in its role, methods, and values that it can be defined as a profession, with conditions of entry and practice regulated by practitioners. However, the benefits of this can be offset by inertia and a lack of response to the needs of change. Friction is far more likely to occur in an area of practice such as design, which is by nature more fluid and less easily defined than other areas. The flexibility and creativity necessary for design practice may be better suited to a less formal definition, but constant problems also arise from this very flexibility, this innate probing of boundaries, that can result in the nature of design practice itself. Since the notion of design is already inherently ill-defined, the scope for misunderstanding and friction is huge. Indeed, the public, in many countries, has long misunderstood what designers do and what their role should be.

FIGURE 14.1 The Evolution of Industrial Design

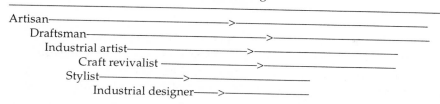

A further point about design needs emphasis. In many respects it is not amenable to simple explanation. Its values and perceptions are not generally understood and must frequently be couched in terms essentially alien to its methods and qualities. In business cultures obsessed with their own narrow procedures and quantitative methods, it can be at a severe disadvantage. In this process by which design is understood, a considerable amount of baggage is inevitably carried from the past. How designers themselves understand what they do, their self-definition, has taken various forms and has progressed through various stages over the last two hundred years; each stage, moreover, has left a legacy as to how they are seen by the wider public. This self-definition often clashes with concepts of what it is believed their role should be in the context in which they work. The resulting confusion has been a source of friction on many occasions in the past and continues to be so today.

In this process of development, although each stage is sequential in time, emergent new stages do not supersede and replace what has gone before. Instead, there is a layering of previous stages as each new one emerges (see Figure 14.1).

A further complexity is that this layering is not a process of simple addition but, rather, one in which any new development alters the relationship between all the different levels. For example, craft work, as a major form of creation and technology, has been widely replaced in many areas of the world by industrial production, but it continues to survive in marginalized forms, such as expensive one-off products for sale in up-market boutiques, or in repetitive parodies of traditional forms for the tourist market. In its original form, however, each stage has characteristics that give it validity.

The Artisan

For much of human history, the processes of shaping and making the objects used in everyday life have been embodied in the traditional craftsman, working in such fields as textiles, pottery, metalworking, woodworking and basketmaking. A feature of such crafts is that the forms used were traditional, handed on to succeeding generations through a process of learning by doing, through emulation rather than precept. This process can be seen, in one sense, as a condition of inertia, and indeed there was considerable social resistance to inventions that precipitated fundamental change. Such resistance can be explained by the fact that in preindustrial society, life was precarious, and in the constant struggle to survive any proven solutions emerging from the collective experience of a social group were valued and not basically questioned. Improvement had to be within the established range of concepts and values. It was a process of accretion rather than step-change.

If the basic condition was one of inertia, it was saved from stagnation by an inversion of the relationship between maker and user from what we consider to be the norm in the modern industrial world. A close, personal relationship generally existed between artisans and users of what they made; indeed, they usually lived in close proximity. In the process of conceiving and making, although traditional concepts of form were not questioned, the artisans were at the same time capable of being infinitely adaptable in detail to the specific needs of any user.

Traditional forms could also, of course, be serially produced by craft techniques if a local market was large enough, and examples in the fields of pottery can be found as far back as 5000 B.C. in Egypt. This serial production was achieved by assembling large numbers of artisans in one place (a manufactory), or through the "putting-out" system, which linked domestic units of production into a coordinated network of distribution and sales. The basic condition could still be regarded as one of inertia in design, however, because the duplication of forms regarded as appropriate to the task did not involve their design being fundamentally altered.

The Draftsman

Two trends vital to the emergence of modern design, and to the possibility of change and friction that has accompanied it, gradually arose over a long period of time in many places: first, in cases where markets increased in size and, above all, in physical distance between producer and user; second, in cases where production techniques began to change on the basis of mechanization and the division of labor. These trends affected how forms were defined and resulted in design becoming increasingly separated into a specialist function.

An example illustrative of the first trend, of increasing distance between producer and user, was the trade between ceramics producers in the Pearl River delta region of China around Guangzhou (Canton) and Western Europe. This trade began with the occasional importation of forms native to China and its culture, often via obscure and indirect routes. By the sixteenth century, however, as trade in these products expanded, Chinese manufacturers not only increased their scale of production but also began to manufacture products specifically for European markets. As a result, draftsmen responsible for drawings of products and decorative forms began to make their appearance in both China and Europe.

Although these draftsmen needed to understand production processes, they were not involved in manufacture. Their role was to adapt forms to European customs that often differed widely from those of China—for example, in terms of the very different shapes used for eating, as in tea or dinner services. In decoration, too, forms were an adaptation of what Chinese draftsmen thought Europeans wanted; they were even produced on the basis of drawings sent out from Europe, depicting what Europeans conceived Chinese decoration should be like.

The friction in this situation was potentially great, but it was avoided because increased production was essentially achieved through duplication of existing techniques and through the capacity for adaptation displayed by Chinese manufacturers. Rather than asserting their own characteristics, which were retained in production for internal use, these manufacturers were willing to bend with the changing winds of taste blowing from Europe if doing so meant a continuity of profitable trade. This tradition of adaptive design can still be seen at work today in the industries and products of Hong Kong and other Southeast Asian countries, and it is one of the region's most abiding characteristics.[2]

The friction resulting from the early stages of industrialization in Europe was much greater on every level, however, because the nature of design and production was fundamentally transformed, with mechanized methods of production resulting in new products that generated demand in new markets, and disruptive social and economic change. With mechanization, the demand for draftsmen capable of specifying products for new methods of industrial manufacture became urgent.

The English ceramics firm founded by Josiah Wedgwood is a case in point. Wedgwood transformed ceramics production on almost every level: by ceaselessly experimenting to find new methods and materials, introducing mechanized methods to increase production, replacing hand methods of shaping forms with standard molds to achieve consistent quantity and quality, and expanding the market with quality products available at affordable prices to a large number of people.

Wedgwood also realized that form and decoration were essential means of adding value to products, requiring that close attention be paid to trends in style. In his time, the cycle of change in fashion was much slower, an important difference, and the neoclassical taste of the period was not only highly appropriate to his products but of much greater duration in public esteem.

Wedgwood understood the role of design in creating new markets, but he had constant problems obtaining competent draftsmen who could adapt to the demands of new methods of production—a concept alien to, and resisted by, traditional craftsmen. Although he engaged William Hackwood, one of the most skilled draftsmen in the trade, and later trained suitable young employees in drawing skills to perform this role, the lack of sufficiently competent designers continued to create difficulties. In what represented a step toward another stage of design practice, Wedgwood turned to artists of repute, employing John Flaxman, Joseph Wright, and George Stubbs, among others, from 1776 onward.[3] Artists were the only people well trained in visual techniques, capable of providing concepts in the form of drawings, sketches, and models that could be freely adapted by the draftsmen and modelers in the factories.

The role of draftsmen as those who prepared detailed drawings for production continued to be one of the main concepts of design practice in Britain well into the twentieth century, for example in locomotive and automobile design. They were often regarded as mere functionaries, but an examination of production drawings from, say, nineteenth-century locomotive design in Britain shows very high standards of skill as well as an ingrained aesthetic sense evident in the proportions of the locomotives and all their details. The term *design draftsmen* is still in widespread use around the world; in contemporary Brazil, for example, it is a common way of describing design practitioners. The Industrial Artist Sir Joshua Reynolds, president of the Royal Academy in London, wrote in one of his discourses in 1776:

> We will take it for granted, that reason is something invariable and fixed in
> the nature of things; and . . . we will conclude, that whatever goes under the

name of taste, which we can fairly bring under the dominion of reason, must be considered as equally exempt from change. . . . The arts would lie open for ever to caprice and casualty, if those who are to judge of their excellencies had no settled principles by which they are to regulate their decisions, and the merit or defect of performances were to be determined by unguided fancy.

Reynolds's belief, albeit in more elaborate and sophisticated form, mirrored the central fact of inertia in craft culture—a deep mistrust of change. Note, however, that the date of his statement coincided with two events that would wholly undermine his worldview: the American Declaration of Independence, and the publication of Adam Smith's seminal treatise on economics, *The Wealth of Nations*.

The point about Reynolds's assertions is that there could have been considerable friction resulting from Wedgwood's use of artists had he too obviously sought to match innovation in form with his revolution in manufacture. Wedgwood's products, however, were highly successful and appealed to a broad middle-class public precisely because they were carefully designed to conform with prevailing taste—indeed, to make it accessible to a wider market. Thus there was no basic clash of principle or interest in his employment of artists who were equally clear as to the role of dominant taste in society. In all aspects, therefore, his success in avoiding friction can also be attributed to adaptability in such matters of taste. The very scale of his technical, manufacturing, and marketing innovations, however, enabled him in other respects to totally subvert those who sought to control it.

That unity of viewpoint in artistic matters did not continue long, as industrialization spread in scope and scale of products and markets, which, together with the democratic impetus they stimulated, swamped the secure standards of taste that Reynolds and other defenders of traditional society and values sought to maintain. By 1830, it was clear that the whole fabric of British society was being fundamentally transformed, affecting much more than just its methods of manufacture, and the country was emerging as the predominant economic power in the world. The industrial leadership of Britain resulted in considerable friction in other countries, where complaints were widespread about the destruction of domestic industries and the swamping of their culture by inexpensive British imports. Many sought to resist this penetration, and the threat to established practice it represented, by erecting protective barriers. Such reactions bear interesting parallels to those against Japanese products in our own time.

Nor, however, was resentment against foreign preeminence wholly absent from Britain as it sought to maintain and extend its industrial dominance. In 1830, Sir Robert Peel, a member of Parliament representing the new class of industrial entrepreneurs, spoke in the House of Commons and fired the opening shots in a campaign to improve British design standards that were inferior to foreign competitors.

By 1836, Parliament had established a Select Committee on Design that, after spending a year considering submissions and evidence, confirmed Peel's views and recommended the establishment of a national system of design education to provide for the needs of British industry. Many industrialists giving evidence seem to have been preoccupied by the inadequate supply of design draftsmen. The new system of education, however, stumbled upon precisely the same obstacle that had earlier thwarted Wedgwood. The only people qualified to teach visual skills and capabilities were academically trained artists, and most design schools therefore became, in essence, art schools. Indeed, the first institution established under this program, and intended to be the model for the system, today bears the title of the Royal College of Art.

Despite fervent efforts to harness art to the service of industry in Victorian Britain, a strong degree of skepticism amongst the art community continued to be expressed. It was all very well for artists to work in industry as long as they could determine the values under which they worked, but such occasions were exceedingly rare. In 1848, a writer in the British journal *Art-Union* stated: "We do not wish artists to become the servants of manufacturers; we do wish them to be their friends and allies; their partners in educating the people; in improving the tastes, and consequently, the morals, of the community; in developing the intellectual strength and the intellectual resources of the United Empire."[4] The disquiet evident in that statement was hardly to be allayed by subsequent developments, but the programmatic advocacy of a moral and educative role for art in industry was to continue and become more highly detailed.

The situation was compounded, however, by the considerable cultural confusion in early industrial Britain, caused by the wrenching changes taking place in society at every level. The products flowing from mechanized industry since the early nineteenth century created new markets and opened the possibility of effective demand, but, it must be remembered, among people who had little previous experience of exercising their judgment. Their criteria for choice were generally inherited from the preindustrial period. In other words, their aspirations were for standards based on craft skill and refined workmanship exercised on decorative forms. The objective was a more genteel way of life.

The new industrial methods could superficially emulate these forms and intricacies of craft culture with increasing ease, but the change, of course, was that once they became widely available in huge numbers, the original conditions of skill and rarity endowing them with value were debased. Moreover, under conditions of commercial competition where added decoration was equated with added value and the possibility of greater profit, a search for distinction and originality to keep the market stimulated led to the indiscriminate use of forms and decoration from any age and any culture. Some of the more technical industries were in a position of compulsory innovation, but other more traditional products had a long history of form and significance that was not easily overcome. In the latter areas, many manufacturers, as Wedgwood had done, turned to the fine arts to engender a sense of distinction and validity in their products. Although prejudice against any tainting of "fine art" with commercial values continued, by the mid-nineteenth century the practice of employing artists to provide decorative schemata for the more expensive ranges of traditional products became widespread and it, too, has continued to this day. It represents a distinct subset of design, usually depending upon "name" designers who cultivate a distinctly individual style in the goods they create. And it is fed by a system of design education that still, to a considerable degree, emphasizes the same role in the graduates it produces. This, in itself, is a major source of friction. A well-known modern example concerns the Frenchman Philippe Starck, whose work is valued for its visual uniqueness rather than for its practical qualities. Chairs designed by him for the Royalton Hotel in New York have three legs with protruding horns from which one may easily fall off when seated or on which one may be impaled when passing by. A toothbrush of Starck's design sells in the United States for $12, four times the price of well-known major brands that probably have far more to recommend them in terms of dental hygiene. To press the claims of utility in such products, however, is to wholly miss the point, for it is clearly neither a high priority for him nor for the people who have the disposable income to afford his creations.

The Designer as Craft Revivalist

Curiously, at the point when the Great Exhibition of 1851 in London had clearly signaled the full extent of British industrial supremacy at the midpoint of the nineteenth century, a major assault on the effects of industrial goods on traditional ways of making and using products began to have great impact on the consciousness of educated opinion. A constant stream of books, essays, and lectures by critics such as John Ruskin and William Morris lamented the breakdown of standards in the lives of both those who produced goods and those who purchased and used them, and heavily influenced the wave of craft revivalism and nostalgia for the preindustrial past embodied in the Arts and Crafts Movement. The appeal of this movement was in stark contrast to the gritty Utilitarianism and reliance of individualism characteristic of industrial philosophy. It drew support predominantly from the English middle class, but since this was precisely the section of society that had been the driving force behind the Industrial Revolution, it represented a serious dilution of the country's capacity to confront the major challenges coming from later entrants to the Industrial Revolution such as the United States and Germany.

Indeed, it is not unreasonable to suggest that such attitudes still continue to hamper attempts to keep Britain abreast of modern technological developments. For example, in recent years, the heir to the throne, Prince Charles, has used his position and authority to advocate a return to an architecture based on images of medieval ruralism and neoclassical style, which by any standards is a curious combination for a major nation on the verge of the twenty-first century. The friction resulting from such attitudes is one of the major subcurrents of contemporary British life.

A recent event provides a fascinating example of this phenomenon and is worth considering in detail. I refer to the replacement of the old bright-red British telephone booth, the results of which aptly illustrate Nordal Åkerman's description of friction as "that which resists, that which is inert and recalcitrant and makes almost every endeavor much harder than anticipated" (again, see Chapter 1).

As part of the Thatcher government's policy to reduce the public sector, the telephone service was separated from the Post Office and privatized under the title of British Telecom (BT). The management of BT decided that its corporate image should reflect the fact that it was now a totally separate concern and, accordingly, in 1987, began to replace the well-known red booths, designed by Sir Giles Gilbert Scott and introduced in 1935, with off-the-shelf booths imported from the United States. These were of plain glass sheet, with black and aluminium trim and an undistinguished logo in yellow printed on the glass. The media storm of protest that hit British Telecom was highly revealing in that it focused almost entirely on the destruction of what was depicted as a "traditional" feature of the English landscape. Nor did the protest quickly blow over.

In 1988, a leading newspaper, the *Sunday Times*, carried an article by columnist Simon Jenkins in which the managers of British Telecom were branded as philistines and the "obliteration from the landscape" of the old booths was described as "a truly national design catastrophe."[5] Interestingly, the name of William Morris was invoked in relation to his thesis that good design is "the expression by man of his pleasure in labour" and that the sight of beautiful things around us "makes our toil happy, our rest fruitful." Exactly how the old red booth achieved these ends was not explained. The black and aluminium BT booths were "visual disasters," stormed Jenkins, with a "phone facia inside that had all the design content of a Russian tank." In contrast, "the Scott kiosks . . . were a milestone in the history of British taste. They were as intrinsic to our image as the Union Jack, the double-decker bus or the Georgian shopfront."[6]

Jenkins's comments were occasioned by an exhibition of designs for a kiosk by a new competitor for BT, the Mercury company, also established as part of the privatization program to provide competition, that he applauded as worthy but in need of improvement. He concluded his article: "But could I plead for a fourth competitor to be allowed on the competition platform? Languishing in Falkirk in Scotland are the original moulds of Scott's magnificent creation. They could well be adapted to the new technology. Until we show we can do better, how about Mercury offering the public a chance to beat a path to that old familiar door?"[7]

In the two years following the new booth's introduction, despite BT Chairman Iain Vallance's citations of independent surveys showing that over 70 percent of customers preferred the new boxes, even shareholders at British Telecom annual meetings continued demands for "a new, designer call box." Reflecting the views of Prince Charles, who opened his campaign against modern architecture by describing a plan for an extension to the National Gallery in London as "a carbuncle," BT shareholders jumped on the bandwagon and also described the new call boxes as "glass carbuncles," demanding "something more like the old red boxes reinstated."[8]. From all these perspectives, it seemed that nothing less than a restoration of the ancien regime in this matter would be satisfactory.

Interestingly, although some complaints against the new boxes concerned their functional qualities, the overwhelming weight of criticism focused on the arbitrary action by a major corporation in destroying what many people regarded as a distinctive and highly characteristic part of British life. Certainly it was a tribute to Gilbert Scott's design that, though only fifty years old, it had come to achieve this status as an integral part of the country's tradition. Nevertheless, the functional deficiencies of the old model, its vulnerability to vandalism, and the problems of cleaning and repairing it were noticeably ignored.

The new model was bland and undistinguished, it is true, but perhaps the most revealing feature of the campaign was the tide of nostalgia and reaction that its introduction released. Missing from public discussions was any attempt either to understand the future of public telephones at a time when the technology was exploding with potential or to determine how a new design could open up possibilities in the areas of both function and visual character. In that sense, the introduction of BT's new telephone booth is an illustration of friction not only in terms of design but of a country's inability to reconcile past and present.

The Designer as Stylist

A very different phase in the evolution of design was associated with the large-scale mass-production industry in the United States, which by 1930 had emerged as the dominant force in the American economy. Henry Ford is often seen as the epitome of this stage of industrial development, with the introduction of his Model T automobile in 1907 and the construction of his Highland Park manufacturing plant in 1913. However, there is an argument, certainly in terms of design, for seeing him as the culmination of nineteenth-century trends, as a summation of the past rather than a pioneer of what was to come. Seen in this light, his career becomes a cautionary tale rather than the example for emulation that Ford wanted it to be.

The reason for this argument is that design for Ford had little meaning beyond basic function. He was, above all, an engineer who believed his Model T to be adequate for all aspects of American social and economic life, simply needing to be produced in ever larger quantities at ever lower cost. It therefore did not need change, refinement, or any emphasis on presentation—these were unnecessary fripperies. A story epitomizing this attitude has to do with his supposed comment that customers could have his automobiles "in any color they like, as long as it's black." Like many apocryphal stories, this one is credible, even if not factually true, since it undoubtedly represented Ford's attitude.

This position could be justified as long as Ford's advantage in terms of production facilities enabled him to dominate the market. Once in a position of leadership, Ford seemed to assume that a rightful state of inertia would prevail, that he could not be challenged. In the early 1920s, however, other manufacturers equipped their plants to standards equal to, and even improving on, those of the Ford company. New conditions of competition, new attitudes toward production and the market, and a new kind of design began to undermine the Model T's dominance. Yet Henry Ford doggedly resisted any suggestion of change to the Model T or to his concept of how the business should be run. In 1927, he took huge losses when forced by collapsing sales to close down plants for several months and hastily reequip for production of a more up-to-date vehicle. But by then it was too late. His company lost its supremacy and never regained it.

What caused Ford's downfall was the emergence of General Motors Corporation as a major competitor in the early 1920s, under the management of Alfred P. Sloan, Jr. Sloan's achievement is an interesting one in the context of friction and inertia. He realized that the system of mass production was, by its nature, basically inflexible—one of inertia. Once production lines were tooled up for the manufacture of a particular model or component, it was difficult and costly to introduce basic change. At the same time, however, he knew that the secret of sales success, under conditions of competition in which there was fundamentally little to distinguish one model from another, lay in providing a compelling image for a potential purchaser. "The consumer recognizes this today," he wrote in his autobiography, "by taking for granted the varied engineering excellence of all the competitive makes of cars, and so his shopping is strongly influenced by variations in style. Automobile design is not, of course, pure fashion, but it is not too much to say that the 'laws' of the Paris dressmakers have come to be a factor in the automobile industry—and woe to the company which ignores them."[9]

The solution he introduced at General Motors (GM) was to generate friction in superficial terms, while adhering to a basic condition of inertia in mechanical components. The tool he used to achieve the former was the annual model change—keeping the market stimulated by constant alterations of form, style, and color, while at the same time deriving maximum economic benefits of scale from long-term production runs of engines, transmissions, and other components. The purpose of "styling," as Sloan termed this design function, was to "create a certain amount of dissatisfaction with past models as compared with new ones." That somewhat delicate description masks the centrality of styling as a tool to produce images that advertising and marketing could use to project a constant impression of change and renewal. "Great skill and artistry are needed to fulfill these complex styling requirements," asserted Sloan.[10] His compliments, however, masked the total subordination of the design function, which lacked any real determining voice, to an essentially executant role in pursuit of marketing strategy.

The success of this strategy was astonishing. It enabled Sloan to build GM into the largest corporation in the world, its economic power exceeding that of most national states. Yet once again, as with Ford, the seeds of decline were sown in the period of greatest success. By the 1970s, General Motors had become a huge bureaucracy, bloated and arrogant, and it was increasingly dominated by concerns about financial manipulation rather than about production of automobiles that people wanted. As a result it was an increasingly easy prey to Japanese and European competitors who emphasized product quality and customer satisfaction. During the 1980s GM lost 10 percent of the total American automobile market, reflecting a huge decline in production, turnover, and profit.

A key factor in this decline was that the manipulation of image had become an end in itself, without any regard to reality in terms of the kind of automobiles manufactured. An example in recent years was the Oldsmobile division of GM. With sales halving between 1986 and 1990, it launched three major advertising campaigns in successive years, which can perhaps be interpreted as a sign of desperation. If another aspect of friction is that it describes the divide between dreams and reality, the experience of Oldsmobile reveals a huge gulf. In the first of its campaigns, the slogan used was "This is not your father's automobile, this is a new generation of Olds." According to an article in *Business Week*, this slogan "drew a lot of attention, but it never explained what an Oldsmobile is supposed to be."[11] The attempt to separate current models from the past in the minds of the public couldn't work because the products were not new or innovative enough to sustain the image being advertised. The solution chosen was, however, also highly indicative—to fire the advertising agency and hire another. The latter's creative talents resulted in an unwarranted level of smugness, given the company's stricken position, as reflected in the slogan "Your gallant men of Olds." This one did not work either, since it was virtually meaningless—and sales still stagnated. The most recent slogan has taken a totally new direction: "The Power of Intelligent Engineering." As a statement and an image it is much more compelling than its predecessors, but to be taken seriously it needs to be backed up by performance. The problem is that although GM's production quality has indeed made welcome improvements of late, there is a legacy of public mistrust from its past performance that will take time to overcome.

The success of Sloan's application of styling in manipulating the public, at a time when the United States constituted a huge, virtually self-sufficient market, led to the imitation of his approach across an enormous spectrum of industries. However, the belief that the manipulation of images, often at huge expense, can become an end in itself and, indeed, can compensate for deficiencies in products is one of the most difficult legacies of the age of styling that still afflicts American industry in its fight to withstand the pressures of global competition.

The Professional Industrial Designer

Since World War II, yet another manifestation of the designer has taken the form of a versatile problemsolver who can be employed by companies either as an in-house designer or as a consultant on the basis of a fee-paying contract. The role of industrial designers in this sense has evolved beyond that of stylists in the comprehensiveness with which they approach a problem; they seek a holistic approach beyond a concern with appearance. Whereas engineers still have a vital role in ensuring that a product works well and safely, industrial designers view their role as being concerned with the point of interaction between product and users, ensuring that users understand the product and are able to apply it appropriately to their needs.

There are two major problems in carrying out this role. One has to do with the difficulties of relationship with professionals in other, longer-established disciplines in industry and commerce. The second problem is grounded in the public's perception of what at any time is appropriate for industrial designers.

The integration of design into the structures of businesses, in ways that make the best use of designers' abilities, is a rarity. Even in one of the more successful examples, Philips, the Dutch multinational electrical products corporation, constant hard work is required to achieve this end. Indeed, Robert Blaich, who for twelve years up to 1991 was managing director of Corporate Industrial Design for Philips, once stated in a personal conversation that it was always easier to integrate design with other disciplines in a newly formed corporate division, where there were fewer preconceptions and no history, than to insert a design function into an existing corporate structure. A major function of managers of design groups, he insisted, is the constant daily task of selling design throughout the corporation.

The problems of friction when design is introduced into existing structures can be observed not only in the inevitable resistance of insiders to the addition of outsiders who may appear to challenge their role or threaten their turf, but also in prevailing perceptions of what design is and the role it should play. Given the range of variations in the evolution of design described here, it is little wonder that there can be serious friction between the way designers see themselves and the way they are perceived by others. If, for example, modern design professionals, thoroughly competent in their field, and capable of understanding and relating to the engineering and business dimensions of their work, are perceived and positioned as artists, friction will be the result. Conversely, given that much of design education continues to produce graduates who view *themselves* as artists, there is an understandable reluctance among people in more established disciplines to take them seriously.

People generally define their needs, naturally enough, in terms of what they already know. They have a remarkable capacity to adapt products to these needs and to their self-perceptions as embodied in the endlessly varying environments they create around themselves. So, when designers attempt to depict themselves as de facto representatives of the public in the development of products, as the final arbiters and determinants of what form and process is appropriate in any situation, they frequently face difficulties in gaining credibility. In moving public acceptance in the direction of what is new, different, or better, they often have to rely upon instinct in the absence of any other measure; and in an age of increasing and often bewildering complexity on every level, that instinct is often inadequate. The issue of understanding users is already causing difficulties as products become more complex; two obvious examples are video recorders and telephones. The telephone has moved from being a single-function instrument to one that can have seventy functions or more, along with possible linkages to other technologies. In all such cases, the potential for friction in a destructive or inhibiting sense is huge.

On another level it is necessary to adapt to the demands and potential of the new age of flexible technology, based on such developments as computers, information systems, and robotics. In the face of fundamental changes now under way, we have to ask; What will the role of design be? No simple answer can suffice, but one important element is already clear. On many levels, the relationship between the form of products and the process of using them, between hardware and software, is undergoing a radical transformation. Changing machine tools, for example, is now much quicker and easier, because in many cases we change not the physical structure of the machine but the information program.

An everyday example is the personal computer. We purchase it not for its physical form, the hardware, but for the process of using it, the software. Software, of course, standardized products to be adapted according to the way we want to use them. In the immediate future, it will possible for us to go a step further and have a modular computer assembled to the configuration and capabilities we want. Indeed, the possibility of adaptation is a crucial difference in many products.

It follows that the nature of design will also be very different. In the era of mass production, the main concern was with style; in the age of flexible technology, the main concern will be the user. The provision of clearer concepts and tools for designers attempting to understand those who use the results of their work can be summarized in terms of the body of knowledge implied by the phrase *human factors*. This phrase originated in ergonomic studies of human dimensions, but it has subsequently been extended into cognitive, social, and cultural disciplines that help explain human reaction to artifacts.

In such a process of change, can it help to look into the past at the very different roles that design in all its manifestations has played? In reviewing the different stages described in this chapter, we find that the only one in which friction was not an innate problem was that of the artisan in traditional society. This exception can be accounted for by the closeness of relationship on every level between maker and user. In all other subsequent stages of evolution, by contrast, the effective separation of designer from user, often by several layers of functionaries in management, marketing, production, distribution, and sales, has been a major generator of friction.

Yet, potentially, that gap can be closed. Both modern production and communications technology are capable of infinite adaptation and variation to the specific and particular needs of individuals or groups. This technology is being inadequately used as yet, due to the innate problems of change on every level, to the inertia represented in established ideas and organizations, and to the inevitable friction involved in moving from one paradigm to another, in the theory and practice not only of design but also of technology, economics, and culture. Design is at the cutting edge in that it is giving shape to this future, whatever form it may take. If friction is an inevitable corollary of the practice of design, understanding it could be essential to minimizing the huge problems of transition that lay ahead and to directing them toward the possibility of creative uses.

PART SEVEN

Physics and Metaphysics

15

An Exemplary Physical Disposition

Rom Harré

Introduction

At first glance the concept of mechanical friction seems not at all odd. It is just another force. It must be considered (as friction) when surfaces slide over one another. It must be taken into account (as viscosity) when motion in liquid media are studied. It must always be included (as resistance) when analyzing motion through the air. But it has certain peculiarities that are most instructive. Provided that two contacting bodies do not move with respect to each other, the friction between them increases with the force tending to make them slide. Physicists call this static or Coulomb dry friction. When the external forces acting on the contacting bodies reach a certain magnitude they begin to slide along one another. But there is still friction. This is dynamic friction. Like static friction it is greater as the pressure holding the two surfaces together increases, but it is always less than static friction in the same case. If we imagine one of the contacting surfaces to be stationary and the other to be moving or tending to move, the force of friction is always opposed to the direction of actual or potential motion.

Newton was the first to realize that friction can largely be treated just like any other force. But in one important respect it is very unlike most other forces. The force of friction appears only in reaction to another force. There is no frictional force between two contacting surfaces unless some outside force "tries" to move them over one another. A pair of surfaces in contact can be said to have a tendency or disposition to resist relative motion. That tendency is displayed as an actual frictional force whenever something tries to push them along. Furthermore, we can never say that just one surface, considered by itself, has such a tendency. Friction comes into existence only in the interaction between two surfaces. Only the *pair* of surfaces has a tendency to resist relative motion.

A student is introduced to physics with some such thought as this: The laws of nature describe how the properties of matter are mutually related and how they change. But what sort of model examples do we have in mind when we think about matter and its properties? A flower and its color? A cat and its ability as a mouser? Most of the examples that spring to mind seem to involve a thing and its properties, as if that thing could have those properties whether or not there were any other things about. But a flower would have no color if there were no people (or bees) to see it, nor could we make sense of the idea of our cat as a hunter unless there were mice to be tracked down. Starting from an analysis of the concept of friction as it used in the physical sciences, I shall propose a general account of material properties, not as attributes of isolated material things, but as relations between and among material things. From the standpoint of that account, "matter," the traditional substance of the world of physics and the usual subject of the ascription of physical properties, will sublimate in favor of a ubiquitous metaphysics of relational properties, both linking and constituting the very things to which they are attributed. I use the image of "sublimation" to emphasize that the concept of "matter" will pass directly from the solid state to invisibility. It will be gone without a trace.

It is convenient to begin this analysis with a pair of concepts later to be discarded. We generally speak of a material substance on the one hand and its physical properties on the other. Philosophers distinguish between individual substances, such as an apple, and mass substances such as water or gold. The science of physics was once built upon the idea that there is a set of absolute physical properties that any material thing, whatever its substance, had of necessity. For example, Galileo listed "having a boundary," "having a shape," "being in motion or at rest," "being in contact or not in contact with some other body," and "being one or many." Of these physical properties, all but contact were thought to be absolute— that is, unaffected by the presence or absence of other material things. As physics evolved in the seventeenth and eighteenth centuries this list was gradually revised in favor of properties that are relational—that in other words, properties exist only in the interaction of one body with others. However, although Newton favored relational concepts in his final account of matter, he took for granted that at least some key properties of matter are absolute. For instance, in the famous thought experiment in which he imagined a pair of globes rotating "in an immense vacuum," he assumed that they would have the same masses as they would have had in the ordinary universe, well stocked with other material things. Mass, for Newton, was an absolute property of matter. Modern physics still makes use of some concepts relating to absolute properties. Students are often invited to imagine physical situations in which just one electron interacts with some given field. In these stories, charge is presented in just the way Newton thought of mass, as an absolute property.

The seventeenth century was also an era of deep investigation into the metaphysical status of the properties that were to be used in physics. What were colors "in the material body" itself? What was heat as it existed in something that felt warm or cold? What was mechanical force? There was plenty of evidence that many perceived qualities of things varied with the state of sensibility of human observers. The same bowl of water would appear warm to a hand that had been in cold water and cold to one that had been in hot water. Some perceived qualities seemed to resemble the physical properties that corresponded to them in reality (primary qualities and the corresponding ideas), whereas others (secondary qualities and the corresponding ideas) did not. Galileo, Locke, and many others explained the distinction between primary and secondary qualities by drawing a logical distinction between occurrent and dispositional properties. Primary qualities were just as we perceived them. Secondary qualities were not as we perceived them. They could only be dispositions. Only ideas of primary qualities were faithful representations of the corresponding occurrent physical state of material bodies. Secondary qualities were physically matched only by powers or dispositions. As Locke put it, a secondary quality is "nothing in the body but the power" to appear colored, to appear hot or cold, to retard the motion of another body in sliding contact, and so on. As to the question "Why does this body appear red and that body yellow?" or "Why do bodies in contact retard each other's motion?" physics can answer only with a description of the occurrent state of the matter of the body or bodies in question. As the physicists of the seventeenth century saw it, this description of the way the colored or rough body really is must be couched in terms of primary qualities. A certain arrangement of atoms selectively reflects light of a certain wavelength. But the question of why a human being should see a yellow surface when stimulated by light of that character was thought to be unanswerable. The program of future physics and its limits were set, so it seemed, once and for all.

Seventeenth-century physicists and philosophers of physics tended to run together the two distinctions I have been making. Those occurrent properties that, as primary qualities, figured centrally in physical science were assumed to be absolute. They were supposed to exist independent of human observers and indeed all other constituents of the physical universe. Dispositional, or secondary, qualities were relational. But they were of minor interest, since they were tied to facts of human sensibility that were unexplainable by physical science.

By the beginning of the eighteenth century a drift toward a general dispositionalism had become easily discernible. However, dispositionalism required a general shift in the kind of state that was supposed to be a manifestation of the disposition. In Locke's account, the manifestations could be either in human sensibility or in the effects on another body. Friction is a special case of the latter. It betokens the move from a concentration on human sensibility to a more general dispositionalism in which the properties taken on by other bodies become the focus of attention. But before considering the development of the modern repertoire of the physical properties of matter, we must rigorously examine the idea of a physical disposition.

A First Account of Dispositions

Philosophers have usually discussed the kind of dispositional predicates that are found widely in chemistry and in the science of materials. "Solubility" and "fragility" have been favorites. There are many simple dispositions ascribed in mechanics as well. Though I shall move on to more complex examples in pursuit of an understanding of the force of friction, some examples of simple dispositions will do excellently to introduce the basic analysis of dispositional predicates that are used to ascribe dispositional properties to material substances. An individual substance, a thing, this cannon ball, has a tendency or disposition to fall if unsupported; we say it *has* weight. An extended substance—say, air—has a disposition to resist compression; we say it *is* elastic. In general the dispositions of individual things, like the air in this bike pump, are treated as instances of the dispositions of the substances of which they are made. But what justifies our saying that the cannon ball, when still lying on the wooden platform, has a tendency to descend? Or that the air in the pump, before the piston is pushed in, is elastic? Dispositions are generally taken to be grounded in some occurrent property or properties of the substance to which they are attributed. The elasticity of air is grounded in its molecular constitution. The tendency of the cannon ball to fall is grounded in its matter.

Unless compressed in a suitable vessel, air does not display elasticity. In all dispositional attributions there is an implicit reference to the conditions under which the disposition is usually displayed *ceteris paribus*. Taking these observations together, we reach the following general format for laying out the explicit and implicit content of a dispositional expression, *D:* "If certain conditions obtain, then a thing or sample of a substance will produce an effect on some other thing by virtue of the nature of the things involved."

In the scientific study of dispositions, this format can serve as a methodological prescription. It raises two questions: What are the conditions under which things and substances react upon each other in well-defined ways? By virtue of what features does a certain thing or substance display, in the appropriate circumstances, this or that disposition? These questions seem to be typical scientific questions. The former could be answered by a simple program of experimental tests. The latter raises more difficult issues. Investigation of the natures of things and substances usually involves deep hypotheses about the existence of unobservable constituents, such as the arrangement of various kinds of atoms in the molecules of the things whose nature is being investigated. It is one of the triumphs of chemistry and metallurgy that ways have been found to investigate just this sort of hypothesis.

Are Dispositions Real?

We typically ascribe dispositions to material substances even when the behavior mentioned in ascribing a disposition is not being displayed. The cannon ball is said to be heavy—that is, to have a disposition to fall, even when it is sitting in its cradle and not falling. Surfaces are said to be rough or smooth, sticky or slippery, even when nothing is in moving contact with them and so there is no net force between the surfaces, no friction.

We can perhaps agree on the following conditions for correctly attributing a disposition to something:

1. A display of the expected behavior has occurred reliably when the relevant conditions have obtained, either in the past history of this thing or with other things of the same natural kind.
2. We have reason to believe that the grounding state persists during those times when the relevant effects are not being displayed. In some particularly central cases the grounding state can serve as a criterion of identity for the substance in question. The fact that water continues to be constituted of fixed proportions of hydrogen and oxygen whether it is freezing or boiling is not only criterial for concluding that it is

water but also serves as the grounding of the disposition to freeze under some conditions and to boil under others.

The question "Are dispositions real?" slips out of focus in favor of the question "Under what conditions are dispositional attributions proper?" In considering the latter question we come to see that the relevant reality issue has to do with the constitutive properties of the substances to which dispositions are ascribed, that is, with the question of the continuous existence of the grounding states of such dispositions.

Varieties of Dispositions

Many kinds of dispositional concepts meet the above conditions. There is variety due to the degree to which the display of a disposition depends on the existence of a particular environment or context for that display. Some dispositions are displayed in a wide variety of contexts, others in very few. Dispositions that display themselves only in humanly contrived conditions for specific human purposes I shall call "affordances." This usage comes from the writings of J. J. Gibson (1968). There is also variety in the extent to which certain disposition are displayed sometimes in one way and sometimes in another. Yet another dimension of variety has to do with the level of activity or passivity of the substance to which the disposition is ascribed. Dispositions that are displayed through the release of the innate activity of the thing or material stuff involved I shall call "powers." For example, an acid has the power to corrode metal. Dispositions that need to be invoked by an outside stimulus I shall call "liabilities." The properties that physicists assign to the things and substances of their *umwelten* are usually affordances. The substances to which they attribute them are frequently taken to be active rather than reactive. Most of the dispositions that physicists study are powers.

By using these concepts I can clarify Niels Bohr's idea of "phenomena" somewhat. According to Bohr, physical properties are unambiguously defined only for specific experimental arrangements. In classical physics a property could be ascribed to a material substance, say, specific gravity to gold, without any reference to the experimental or observational setup in which specific gravity is to be displayed. That leaves the application and meaning of the concept vague. In classic physics this deep-seated ambiguity in the meaning of physical properties is of little practical importance since they manifest themselves in much the same kind of display in every relevant context. At least so it seems. Even such a context-dependent interactional disposition as friction seems to be unproblematic. However, the polarity of a magnetic field is displayed in one experimental context in the phenomenon of attraction/repulsion but in another in the phenomenon of electromagnetic induction of a current in a conductor. Classic physicists tended to see these varieties as merely contingent variations on a uniform fundamental disposition. The magnetic field had polarity whether it was interacting with another such field or with the electrons in a conductor.

Bohr realized that if the relevant disposition was displayed only in the interaction of the stuff of the world with a particular kind of experimental setup, then it was, as we would now say, an affordance of that setup. The world stuff did indeed have that disposition, but it was to afford this reaction in just this experimental arrangement, and not in some other. So the concept was determinate only for that experimental setup. The characteristics of one experimental arrangement might very well preclude the setting up of another with different characteristics simultaneously and in the same place. In that case what would be afforded by the second could never be achieved during the running of the first. If momentum was afforded in one setup and position in another setup was excluded by the conditions for setting up the first experiment, then the things in the world do not *have* either momentum or position. To suppose that they did would be to imagine that there were physical properties that existed independent of the apparatus through which they became manifest to an experimenter. Bohr's idea was that these apparently objective properties were affordances, displayed in alternative but complementary pairs in pieces of equipment whose design precluded their being run simultaneously and in the same place. We should not ascribe the responsibility for this restriction to weird properties of subatomic entities. There are no such entities in the absence of the working apparatus. There are only natural dispositions to afford such entities in the apparatus. It is impossible to build a detecting screen that is at the same moment both fixed (for affording particulate position) and movable (for affording particulate momentum). Only if we mistakenly think that the world includes entities that have the properties displayed in the experimental interaction designed by the human beings who study them do we get the apparent paradoxes and absurdities that seem to appear when we try to match concepts drawn from common sense and from the older physics with the phenomena of the new. To pursue this issue further, we would need to investigate subtle questions about the nature of models used by physicists, but this would take us beyond the scope of the chapter.

The dimension of activity/passivity is of no less importance for understanding the significance of the concepts of physics. The focus will now be on the entities and substances to which the affordances are ascribed, rather than on the display of those affordances in human experience or in the reactions of instruments. The story of the ups and downs of activity concepts can be plotted as a kind of war between two groups of philosophers. There were those, like Hume, who mounted critical attacks on the use of such concepts as "active power." And there were those, like Kant, who were sensitive to the fact of the persistent reintroduction of such concepts by physicists faced with the problem of describing and accounting for their experimental and observational results. One way of bringing to light what the arguments have been about is through an examination of some schematic explanations of mechanical and electromagnetic phenomena. Let's say that some identifiable and bounded material thing moves, relative to some fixed frame of other things, to which a convenient coordinate system has been "bolted." Why did it move? I chose the above form of words advisedly. By referring to motion relative to a fixed frame I preclude the flip answer "It didn't move, the frame did!" (There are complications when we try to fit this simple picture into the context of special relativity.) Since antiquity, physicists have had two explanatory schemata available.

The passivity schema refers the source of the motion to an extrinsic influence. In the heyday of the mechanical world picture, the effective influence was always supposed to be exerted by another material thing, moving relative to the frame, and colliding with the passive entity whose subsequent motion is thus explained (*ceteris paribus*). The concepts of "momentum" and "energy" permitted the whole cause of the mobility of the impacted entity to be ascribed to that which impacted upon it. Only the inertia of the former was drawn into the explanation of the motion through the need to account for the different degrees of motion suffered by different bodies when subjected to the same extrinsic conditions. Newton's famous laws of impact enshrine just this schema.

Historians have often remarked on Newton's struggles to accommodate gravitational phenomena within a universal generalization of the classical schema. In Query 31, appended to his *Optics* of 1704, Newton plays with a bouquet of hypotheses that would permit the intrinsic tendency of a pair of material bodies to attract one another to be treated as the effect of an extrinsic influence exercised through contact between them. The conception of gravitational "action at a distance" that Newton tried to exorcise is just one application of the activity schema of explanation. Motion, in some cases, comes about because the body under scrutiny, while still stationary, has an intrinsic tendency to move but its manifestation is prevented by some block. When the block is removed and the tendency released, the body will move. Why introduce the tendency? Isn't it enough to correlate the removal of the block with the motion? A material body falls in a gravitational field because whatever was supporting it was removed. Must we include in the repertoire of beings acknowledged by the great science of physics both plain, robust material things and mysterious, occult qualities, tendencies, dispositions, and powers?

It was to this debate that Hume brought his famous skeptical analysis of the idea of active causation. Hume's argument against the use of dispositional concepts in the language of physics rests on a certain technique for ascertaining the meaning of a concept. To find a meaning, one tries to identify the impression that corresponds to it. The term *impression* had a special meaning in Hume's philosophy. An impression was an atom of human experience. Hume argued that there were no impressions among the elementary units of one's experiences of material things and events that could give meaning to dispositional concepts. In a famous passage he declared that when a human being perceives a causal interaction, the only impressions constitutive of that experience are those of the prior event or state and the subsequent event or state. The causation, the productive relation between them, is not perceived, so there can be no impression of it. Yet the idea or concept of causation includes an idea of productive efficacy or power. The frictional force, we say, eventually stops the cart rolling. How can that make any sense? "Cherchez l'impression!" Hume finds the relevant impression in the psychological phenomenon of the expectation induced in a human being by repeated observations of a regularity between similar pairs of events. Degenerate versions of this powerful argument have survived to the present day. For example, I have heard it objected that one cannot use dispositional concepts to describe the material world because one cannot perceive possibilities. However, at least as far as the philosophy of physics is concerned, the Humean objection need not trouble us. Dispositions, as affordances and powers, are ascribed to imperceptible substances. The experimental techniques of physical science are concerned to produce perceptible responses in material things. I happen to think that causal efficacy is perceptible (Harré and Madden, 1977), but the considerations that can be raised for and against that claim need not be rehearsed here. Rather, it is the viability of the use of dispositional concepts in the discourse of physics that is at issue.

There is a further twist to this analysis, first clearly presented by Newton. It can also be illustrated with the example of friction. When two surfaces are sliding one over the other, there is then and there a force of friction. It is an active power at work. The disposition that we ascribe to two surfaces of the type with which we are experimenting must be grounded in some permanent state of the surfaces. If there is a variable force, says Newton, we must suppose there to be an "agent acting constantly," some invariant, permanent state of the physical setup. The variation in the actual force is to be accounted for by variation in the conditions. In the case of friction, the invariant aspect of the setup of surfaces in contact is their surface texture. The variable element lies in the conditions. In this case it will be the force with which they are pressed together and the forces with which they are driven to slide or to tend to slide along one another. Thus we arrive at a second account of dispositions.

The Second Account of Dispositions

A number of unexamined assumptions infect the previous analysis. In particular, the role of substances in the grounding of dispositions needs careful examination. Physics is committed to a structural account of material stuff. It might have been otherwise. By the beginning of the seventeenth century there were two main contenders for a metaphysics of material stuff. Revisions and revivals of Aristotelian/Hermetic theories of matter led to an account based on the idea of "principles." There were several versions of this scheme. In the longest-running theory the "four principles" were "the hot," "the cold," "the wet," and "the dry." The qualities of earthy substances, for example, were derived from their proportions of the cold and the dry. There were also "three principle" versions about. One influential scheme was proposed by Paracelsus. He identified the chemical "elements" mercury, sulphur, and salt as the bearers of the three basic principles of matter. The idea of principles persisted into the modern age. For instance, Lamarck used such a theory in his chemical work. However, by the beginning of the nineteenth century the victory of the atomic/ structural conception of matter was almost complete. The history of a contest is usually written by the victors. The alternative metaphysical scheme, if anyone thought about it at all, was discarded as the rubbishy emanation of overheated imaginations. Paracelsus and Lamarck (as chemist) retreat to the margins of the pantheon of chemistry and physics, of which the central plinth was occupied by such worthies as Boyle, Lavoisier, and Frankland.

In the twentieth century, however, physics has evolved in ways that echo some of the pre-atomist ideas. We may have come to the end of the atomic/structural era. I shall set out the basic ideas of the rival schemes and examine them with respect to their ability to deal with the problems raised by the ubiquity of dispositional concepts. Friction, seemingly such a simple phenomenon, will serve as the paradigmatic example. Aristotle's doctrine of the four elements seems to have anticipated some of the metaphysical foundations required by contemporary physics.

The four elements are usually rendered in English as "earth," "air," "fire," and "water," though they are not to be identified with these stuffs as we know them in ordinary mundane contexts. What is the relation of the four "elements" to the four basic qualities, "hot," "cold," "wet," and "dry"? My observations here are closely modelled on R. Sokolowski's (1970) treatment of the Aristotelian doctrines of elements, substances, and qualities. The "official" interpretation, as crystallized in medieval interpretations of Aristotle, required that the elements be substances, in the true sense, having substantial forms. But Sokolowski shows that there is good textual evidence for ascribing a more radical and interesting opinion to Aristotle, namely that the elements are "substances only potentially." They do not have substantial forms. Sokolowski's reading turns on the relation in Aristotle's general physics between the qualities and the elements. The main evidence for the radical reading comes from *De generatione et corruptione*. According to Sokolowski, Aristotle's derivation of the doctrine of the four elements rests on an analysis of the possible combinations of the four qualities. But these qualities are not current properties of things. They are dispositions. There are a pair of liabilities or capacities and a pair of powers or activities. "Solidity" and "fluidity" are "capacities to *receive* pressures," the one resisting and the other yielding, though both maintain intact boundaries. "Heat" and "cold" are abilities "to *act*" on other bodies. Heat fuses the homogeneous and separates the heterogeneous, whereas cold congeals everything, regardless of kind.

How are the qualities (powers) and the elements (substances) related? In the classic view the qualities are powers of the elementary substances. Sokolowski points out that elsewhere in *De generatione et corruptione* (II/1, 329a, 32–35), Aristotle clearly seems to take the qualities as constitutive of the forms of the elements rather than "following from them" (Sokolowski, 1970, p. 268). Further textual evidence for the ontological primacy of powers in Aristotle's general physics can be found in the same text—for instance, the use of the expression *so-called elements* for the classic foursome, the shift from use of the word *elements* to identify the powers as constituting the substantial fundamentals when Aristotle turns from expounding the opinions of others to outlining his own view, and so on.

The atomic/structural constituent metaphysical scheme is essentially hierarchical. The affordances and powers of material beings are grounded in their constituent "atomic" structures. "Heat," as a complex dispositional property, has been described for more than a century as being grounded in the motion of the molecular constituents of hot and cold substances. Explanatory formats like this work by taking for granted certain properties of the "atomic" constituents of the material stuffs in question. The program of physics since the late nineteenth century has been animated by the project of repeating, in an appropriate form, the program that has proved so successful in chemistry. Each time the scientific community reaches a certain level of analysis—say, in terms of the powers of the atoms of chemical elements—a new structure/constituent explanation is set up through hypotheses about a new kind of constituent of matter. The structural organization and intrinsic properties of the new class of constituents provide the grounding for ascribing those dispositions previously ascribed to the chemical atoms of the old level of deep analysis. When the term *atom* was appropriated by chemists for their unit of structure, the field of subatomic science was born. But the animating metaphysical principle of that field is just the structure/constituent scheme reapplied to the problem of understanding the affordances and powers of whatever was elementary at the former level. So far as I can see, it might have been otherwise. The next step in physics might have been animated by a reintroduction of the Paracelsan metaphysics of principles. The fact that the structure/constituent scheme could be reapplied to build subatomic physics does not seem to me to establish that it must be repeated yet again to build a subsubatomic physics. I shall try to show that the tide of physical theory and experiment has moved toward a situation in which the chain of applications of the structure/constituent metaphysics may have come to an end.

Even without the march of history to suggest a change of tune, doubts about the ubiquity of the structure/constituent metaphysics are easily raised. The progress of physics can be represented according to that scheme as a sequence of steps, each of which repeats the form of the one immediately preceding. The powers of surfaces to resist sliding can be referred to their physical nature, which is explicated as a structure of atoms. The constituent atoms are themselves assigned certain powers. These powers can be referred to the nature of the atoms. This can be done if we conceive of them, in their turn, as structures of subatomic particles. The constituent subatomic particles are assigned dispositional properties such as charge and spin. These are then referred to yet another level of structures of constituents, the quarks and the exchange "particles" that carry the relevant binding forces. And so on. It is in contemplating the "and so on" that the troubles with the powers/natures hierarchy of hierarchies (with "natures" explicated through the structure/constituent metaphysics) come to the surface.

Faced with a regress, a philosopher typically asks whether it is open or closed. Let us ask this question of the regress we have exposed in analyzing the physicists' account of so commonplace a phenomenon as friction. If the hierarchy of structures is open, is the human *umwelt* destined to expand indefinitely just so long as there are physicists to explore it and indulgent governments to pay for their adventures? Or is there an ultimate level of constituents that have no parts, so to speak? Neither alternative seems welcome. Let us look more closely at both of them.

The idea that the universe should be infinitely complex does seem intuitively implausible. But to give a rational form to that intuition proves very difficult. Kant's antinomies purport to show that with the help of reason we human beings cannot draw a conclusion about either the infinity or the finitude of the totality of material beings (things and events) in space or in time. The history of the physical sciences yields a clear picture of a progression from powers to natures, and yet only the first three or four steps in that progression have so far been taken. Poor support for the hypothesis that indefinitely many such steps lie in front of the physics community in the coming millennia!

I can see no way ahead in the project of finding evidence or arguments for or against the hypothesis of infinite complexity. So I shall turn my attention to the hypothesis of finitude. All the occurrent properties of material beings that are revealed by observation or ascribed through the matching of models in theorizing are treated by the science of physics as displays of dispositions. Within the framework of physics no finite regress of powers and natures could end in a substance to which some set of essential occurrent properties is ascribed. Whatever attributes are ascribed to the ultimate beings must include some that are definitive of the kind to which the being belong. By hypothesis there is no deeper level of "atomic" structure that could serve the role of kind determining. At any higher level all the observed or manipulated attributes might be thought to be accidental, environmentally induced, on some basis of unobservable and theoretically imagined necessary attributes.

The "last" ascriptions must be ascriptions of dispositions, but to a substance whose nature is exhausted by those very dispositions. The model for a logic of ascription throughout the descending or ascending regress of structure/constituent statements has been the subject and predicate form, the grammatical realization of the metaphysical pair "substance/attribute." What is the substance of which the "last" dispositions are attributes? *A fortiori*, it can have no other attributes than those very dispositions. The world structure must finally be grounded, it seems, in a substance with no "inner" nature. But the motivating principle through which the regress was built was that every disposition is grounded in some aspect of the nature of the substance to which it is attributed other than that very disposition, on pain of falling into the trap exemplified by the *"virtus dormitiva."* Our grounding substance has no properties other than those basic dispositions. If there are ungrounded dispositions at the basis of physics, why should they not figure at every level, including that of human perception? This is essentially the objection raised by J. L. Mackie (1973) to the rampant dispositionalism of those who, like myself, advocate a metaphysics for contemporary physics based on the concept of a causal power.

There are escape routes from this seeming inconsistency in the foundations. One route involves the claim that there must be a qualityless substance. It would have to be an "almost nothing" defined only by the one disposition, namely, a disposition to take on the dispositions on which physics shows the material world to be grounded. This is the route that we have already traversed with Aristotle, guided by Sokolowski as exegeticist. The concept of friction can once again serve as an example to illustrate the way the "substance" energy can be usefully invoked. Friction became important in the nineteenth century as the phenomenon best suited for the experimental investigation of the transformations of energy from one form to another. In every physical phenomenon, energetists imagined that an active entity was involved: the energy. There was the kinetic energy of motion, the heat energy absorbed and released in heating and cooling, the electrical energy displayed in lightning bolts, and so on. Could they be transformed into one another? Or to put the matter another way, was there just one "energy substance" that appeared in a variety of guises in different physical situations? In the phenomenon of friction there was a simple and readily researchable case of one form of energy, namely mechanical, being transformed into another form, namely heat. Joule devised a simple but powerful experiment to measure the proportionality between the two forms. He set up a mechanically driven paddle wheel in an insulated jar of water. The friction between the paddles and the liquid appeared as heat. The temperature change in the water for a given mechanical action could be measured and the constant of equivalence calculated. Of course, this program of experiments made sense only in the framework of a metaphysical theory, namely, that all the various dispositions displayed in the phenomena studied are grounded in one ubiquitous substance, energy. It is the route that is presently represented by the concept of "energy."

There remains yet another way out. Dispositions could be raised to the status of basic particulars. We could make dispositions themselves the basic entities of the physical world. Objections to such a radical proposal usually take the form of complaints that this move would ask us to think of an occurrent world, a world existing here and now, that from which our current *umwelt* has been carved out, consisting of mere possibilities. How could an actual world be constituted of possibilities? But we could say that dispositionalism does not require one to believe that an actual world is made of mere possibilities. Rather, *we* are confined to certain forms of description when we want to say something about dispositions. In particular we are obliged to speak and write in terms of possibilities—that is, in the "if . . . then . . ." form.

But whatever constraints there may be along the way, we have to construct descriptions of powers out of predicates that refer conditionally to the conditions in which they could act and to the effects they would produce; in short, we have to describe continuously existing centers of activity. The possibilities through which dispositions as affordances are described are relational properties in which the power as such stands to human beings or to some of their equipment. The modality of the existence of powers as basic particulars at specific places and times is not possibility *de re*. I am not saying that possibilities are some strange kind of thing.

Friction seems so commonplace a phenomenon and so much on the surface of things! And yet a careful study of its place in the history of physics has shown us how it stands on the boundary between the observable and the unobservable, a boundary that has been the subject of so much deep philosophical investigation and argument. The way physicists have treated the phenomenon of sliding resistance, its causes and its effects, illustrates the deepest metaphysical problems of the physical sciences—in particular, the enigma of the status of those dispositions that we must invoke in explaining what seem to be the simplest and most ordinary of physical processes.

References

Bohr, N. *Atomic Physics and Human Knowledge.* New York: Science Editions, 1958.
Gibson, J. J. *The Senses Considered as Perceptual Systems.* London: Georg Allen and Unwin, 1968.
Harré, R, and Madden, E. H. *Causal powers.* Oxford: Blackwell, 1977.
Hume, D. *A Treatise for Human Understanding* (1739). London: Fontana edition, 1962.
Mackie, J. L. *Truth, Probability and Paradox.* Oxford: Clarendon Press, 1973.
Newton, I. *Optics* (1704). New York: Dover edition, 1952.
Sokolowski, R. *The Formation of Husserl's Concept of Constitution.* The Hague, 1970.

16

Friction of Bodies, Friction of Minds

Agnes Heller

I

In the middle of the fifteenth century, roughly three hundred years before Newton formulated the so-called laws of motion, Nicholas of Cusa, the philosopher-cardinal, had pondered the phenomenon that has been known since as "rolling friction." He said as follows:

> Notice that the movement of the ball declines and ceases, leaving the ball sound and whole, because the motion that is within the ball is not natural, but accidental and violent. Therefore when the impetus that is impressed upon it dies out, it stops. But if that ball were perfectly round...its motion would be round. That motion would be natural and in no way violent, and would never cease.

Certainly, nothing can be completely round—thus all bodies suffer friction and wear. "Only the intellectual motion of the human soul, which exists and functions without the body, does not cease" (De Ludo Globi, Vol. CLV).

Cusanus did not distinguish the observation and explanation of natural phenomena from the metaphysical meaning or interpretation of the same phenomena. His interest in observation was thoroughly intertwined with his metaphysical speculation. Cusanus is succinct in a scientific fashion. He attributes rolling friction (and wear) only to inorganic bodies, because their motion is not substantial. Once the motion is substantial (in organic bodies), there is no wear, and the cessation of the movement cannot be explained merely in terms of mechanics. The physical observation stands for the metaphysical truth concerning the substantial propensity of the soul. The ball game itself, the occasion for Cusa's physical observation, stands for many a thing besides itself. Men and women are like balls, they are thrown onto the surface of the earth. Each and every (human) ball is different (none is completely round); and different as well is their original impetus—they will not stop at the same time. In addition, as intellects endowed with substantial movement, humans can change the curve of their movement on the earthly surface; thus they can stop closer to, or farther from, the middle of the circle of the (symbolic) earthly spheres, which is the point of grace. For Cusa, the whole world and all the phenomena of the world were merely finite symbols serving the approach to the Infinite, which can never be entirely comprehended.

Three hundred years later, physics and metaphysics are still intertwined in Leibniz. True, he makes frequent references to the difference between the physical and the metaphysical point of view. For example, so he says, it makes no sense to ask metaphysically whether the boat causes the rings on the water or the water causes them, given that causation is entirely relative to the standpoint of God (who causes the total phenomenon); yet in physics, the question can be answered. But the two aspects remain in concert. This is perhaps nowhere as obvious as in the philosopher's radical and spiteful rejection of the Newtonian theory of gravitation. The idea that bodies can affect each other without touching each other—that is, without a kind of friction—was in his mind a horrendous one. Vacuum, so Leibniz argues (in "A Specimen of Discoveries About Marvellous Secrets"), is inconsistent with the perfection of things, since it interrupts the communication of bodies and the mutual strife of all with all. And only in the case of universal communication within the great chain of beings can every single and unique body (and soul) affect, if only remotely, all others and vice versa. Hegel's and Goethe's hostility toward Newton was in turn motivated by the premonition of the demise of a unified, and also organic, vision of the world order, with or without a metaphysical foundation.

As long as the observation of physico-biological phenomena remains intertwined with metaphysics, the *names* that describe those physical phenomena have *no metaphoric use*. Metaphoric use has to be succinctly distinguished from literary use. If a name that refers to a natural phenomenon, a mathematical equation, a geometrical figure, or the like entails a symbolic dimension, that name does not lend itself to a mere literary reading and, hence, does not lend itself to a metaphoric reading either. The referent itself is multidimensional, because it stands not only for different shades of a meaning but also for a variety of meanings that unfold on entirely different levels, within different spheres of discourse. These names are shorthand versions of a text, an interpretandum; that is, they are not simply referred to but also alluded to. For example, the word *sphere* that I just used is pregnant with a multiplicity of overlapping and nonoverlapping meanings; actually, each author reshapes these meanings. The same can be said of concepts such as roundness, impetus, force, attraction, harmony, sympathy, and even of the Circle, the One, the Ten, the Infinite, and the like.

We have learned from Kuhn that paradigms (in natural sciences), and thus natural sciences as such, appear in the postmetaphysical era. Yet we have also read in Blumenberg's *The Legitimation of the Modern Age* that Newton's vision of the physical universe is metaphysically founded. There is no real contradiction between the two assertions. Metaphysical foundation, embeddedness, is one thing, and the undistinguishable physical/metaphysical (symbolic) message of observation sentences is another. The latter is not the case in Newton and in the natural sciences after Newton. This is why Weber could discuss the "disenchantment of the world" by modern sciences.

It is in this process of disenchantment that terms, categories, and laws of mechanics (and of physics in general) as well as those of chemistry and biology have assumed metaphoric use. Our everyday language abounds in such metaphors. A newly observed or described natural phenomenon has to be "baptized," and to that purpose natural scientists usually borrow a name from everyday parlance by analogy. For example, the words *resistance* and *transference* have been borrowed by physics and by psychoanalysis alike. Almost all of these words are then reclaimed, with a vengeance, by everyday speakers or by writers and poets; they assume a thin or thick metaphoric content. Think of the title, *Elective Affinities*, that Goethe gave to his novel on love-chemistry, or of the everyday use of terms such as *energy* or the *big bang*.

II

The story of "friction" has yet to be written.

Certainly, everyone learns about friction in high school. And if we think over what we learned, we will come to understand that friction is one of the major factors of the human condition. Without friction we could not walk or grasp. Without friction no thing that has ever been created would ever be used up. Aristotle's book *Generation and Corruption* would be in need of rewriting—but, alas, without friction no one could write.

If we want to know a little more about friction, we can look in an encyclopedia. Here we will learn that friction belongs, together with wear and lubrication, to the tribological phenomena: "In almost all cases, the friction force F, defined as the force required to produce or maintain sliding, is proportional to a force L which is normal to the surface, so that if a force smaller than F is applied to stationary surfaces, no sliding occurs. The constant of proportionality, equal to the ratio of the two forces F/L, is called the coefficient of friction F." In the case of a surface called Yugoslavia, we could ponder what kind of (armed) force is required to "produce or maintain sliding" of the state into nothingness. Or when we read about measurable transfer in a case when solid bodies touch each other, we can wonder about the measurable transfer of libido (energy?) or of anger, frustration or dirt, depending on the kinds of bodies (although not quite solid ones) that happen to touch each other, as well as on the circumstances of their encounter.

All three terms that refer to one of the three "tribological phenomena" are used by physicists in a literal way. None of the terms (as terms) is deeper or more important than the other two. They simply refer to different phenomena, or to other aspects of the same physical constellation. Say *wear*, say *friction*, say *lubrication*—the physicist, and particularly the expert of the tribological phenomena, will know exactly what you mean. However, the same cannot be said about laypersons. They might not associate anything with the word *lubrication*. They could say a lot about *wear and tear*, of course, but with very little metaphorical content. However, friction is quite different. The word *friction* is rich in metaphorical content and associations. Why friction, rather than wear or lubrication? The successful career of a metaphor cannot be explained; or, rather, too many explanations can be found for it. Reflecting on the variety of the uses of the metaphor might offer certain clues.

III

Habeas corpus, the inviolability of the human body, is a fundamental creed of modern civilization. But to determine when, and how, and to what extent a human body has been violated is not easy, for apart from extreme cases, such as killing, injuring, and imprisoning, the perception of harm done or of the unharmed state preserved varies from culture to culture. Civilizations develop highly sophisticated kinds of body language. How another human body—or an organic/inorganic body belonging to another person—can be touched, and when, is an extremely important variable in such a language. Since human movements are supposed to be voluntary, infringements of the code of body language of a civilization are offensive, unless they are neutralized by being declared involuntary.

Children sometimes beat lifeless things if they cause harm. Adults often do the opposite. When they push the glass and spill the wine of another person involuntarily, they might apologize with an "excuse me" that roughly means "Regard me not as a living and conscious actor but as if I were a lifeless puppet." "I am sorry" and "pardon me" can be used as equivalents, although they have a stronger appeal by including a reference to the living soul of the offender who begs for pardon or might be sorry. The limit to bodily contact (pushing, patting, touching, pulling) is a variable according to civilizations, social standing, sex, age, or intimacy. Where social arrangements are asymmetrical, so, by definition, are the limits set for bodily contact. It needs to be stressed that even in modern times, after the basic social arrangements became organized along the patterns of symmetric reciprocity, body language still preserved major patterns of asymmetry. This is obvious in the relations between the sexes and between adults and children. Moreover, in children's relations to each other the "archaic" pattern frequently prevails. *The Lord of the Flies* is an allegory that makes this point.

The arrangement of human bodies in space is quite complex, particularly so in modern civilizations where workplace and home are divided, multigenerational families are increasingly rare, and life in big cities becomes a matter of course. The symbol of the circle can hardly do justice to the contemporary perception of what is close and what is, or becomes, remote. My "closest relatives" can live at the other end of the world, whereas I sit in the same small office together with the same half-strangers every day, and I am pushed on the subway by total strangers.

The multiplicity of experiencing "closeness" and "remoteness" today results from the dissolution of a more traditional way of life. Actually, in several places of the world, particularly in villages, the "circle model" still prevails. Closeness/remoteness is a binary much like occasional/steady, conventional/chosen, partial/total, influential/noninfluential, and warm/cold. In the traditional world "steady," "total," "influential," and "warm" normally went together, and "conventional/chosen" remained the most important variable, although the "chosen one" was normally also chosen from the inner "circle" of friends and acquaintances, as happens in Thornton Wilder's play, *My Town*.

The influential/noninfluential binary deserves some attention here, for it prompts us to reappraise the Leibniz-Newton controversy. Newton provoked his age by coming up with a world-explanation in which bodies that do not touch each other can be attracted to, and repulsed by, each other. That bodies unknown to each other can influence each other's life across a certain "vacuum" has become by now an elementary experience. The body called Joseph Stalin changed the life of hundreds of millions of people more than anyone else in their close environment; so did his corpse.

But the intervention of remote bodies into anybody's or everybody's or somebody's life (attraction and repulsion, in a less metaphorical interpretation, included) has little to do with the metaphor of friction. Friction arises in situations where the constant closeness of bodies becomes an emotional strain for one of the parties involved or for all of them. Where there is no warmth or closeness of any kind, there will be no friction.

Warmth or closeness are certainly relative terms. The closeness among the members of a string quartet is different from the closeness of parents and children; but there is closeness in both relations. Similarly, there can be closeness among business partners if they are dedicated to a common (business) purpose. Warmth need not be that of love, it can also be that of respect, recognition, or simply of habit (one likes to see the same face around oneself). Warmth can be mentioned if one misses the other person who has decided to leave (the family home, the business, the string quartet).

Friction can occur not only among individuals but also among groups that occupy a certain space in corporate bodies (e.g., communities, parties, trade unions, and the like; sometimes even confederations with a long history behind them). One can discuss friction among different groups in a kibbutz or among factions of a party.

Friction need not end in breaking away or breaking up, it can be healed. Like *friction*, the expressions *breaking up* and *breaking away* have equivalents in the science of physics. Healing is different. Yet we know that living bodies do not suffer wear through (physical) friction, because the affected organic tissue regrows. The same can happen in the case of metaphorical frictions.

IV

Nicholas of Cusa compared humans with balls, or rather with spheres, which once thrown upon the surface of the world will stop at different times and in different spots depending mostly on the impetus of the throw. No two balls, so Cusanus insists, will stop at the same place, for each of them is different and unique. In the ball game invented by Cusanus, balls did not strive to occupy the same place, and nothing was said about the eventuality of one ball hitting the other. Similarly, in Leibniz's world, monads could not collide with one another given that their commerce was guided by universal harmony. In the mystical/metaphysical vision of the universe there is accord, not discord. The organic and the inorganic are merely stages of the same continuum.

Since then, Newton's ideas have transformed physical nature into an aggregate of dead bodies without purpose. Philosophy's renewed attempts to save the unity of the Cosmos have become marginalized. As a consequence, a new science, biology, has emerged and conquered. Biological imagination is neither mechanical nor mathematico-spiritual. Images of the survival of the fittest (body or mind), of the struggle for life, dominate this biological imagination. Certainly, biological imagination does accompany the development of biological science—it begins to live a relatively independent life in and through metaphors. Friction (of solid bodies) is a force, but not a life-force. The metaphors of force and the metaphors of life-force are different in kind. Life-force is a mythological creature; so is the force of death (the Angel of Death). Now they are resurrected as Eros and Thanatos, respectively.

The body as it appears in biological imagination does not touch and push and resist and pull; nor does it slide or roll or sit. Rather, it desires, catches, devours, conquers, suffers, or submits. As long as physics was clad in the robes of metaphysics, force was mystical; it had something to do with circles and numbers, with the sun and the light and the heat, and so with good and evil. But when those circles and numbers, those surfaces, slidings, the sun and the fire, the light and the throw lost their symbolic meaning, the mystic dimension of "forces" evaporated and the mythology of life-forces occupied the vacant place. The most frequently used metaphors are Sex (or Gender), Race, Life as Survival, Drive, Instinct, Cancer (as Susan Sontag saw it), Growth, Health, and Pollution.

Friction is a funny metaphor, for it is outdated, melancholy, and also utopian. There is nothing mythological about friction; it would be pathetic to render the word *friction* with a capital *F*, though we have no second thoughts about rendering all the biomythological metaphors with capital letters. *Friction* is also a minor metaphor of limited use. This becomes immediately clear if we compare it with its brother *force* or its cousin *life-force*.

One may wonder whether *friction* is an authentic metaphor at all; but it certainly is. The term means literally a force that resists the sliding of one solid object over another (a matter of kinetic friction, rolling friction, or static friction.) However, the metaphor *friction* refers to souls or minds, although it is not unrelated to the relation of bodies. Friction occurs when bodies and minds (souls) are close, and when conflicts (or embryonic conflicts) occur because of this closeness. If there is friction, both the mind (soul) and the body of the persons are involved in the friction, regardless of whether it occurs in a family or among friends, business partners, or members of the same string quartet. This is so because both physical closeness and emotional closeness are presupposed in friction. The same can be said about the friction that arises among groups of collective or corporate bodies. There is no friction without a kind of unity, no friction without a kind of emotional involvement in this unity, yet also no friction without both unity and the emotional involvement becoming problematic. There is an act of challenge or rebellion in friction, but not an outright opposition. There is suffering too, or at least discomfort. This is why friction is a melancholy event, or rather a melancholy metaphor. Friction is an outdated metaphor, because unlike force, energy, or power, it is unfit for being extended to the field of biological imagination. Yet it preserves certain mystical elements of the old metaphysical tradition. As I suggested, where there is friction, there must also be an original unity, and friction is the moment when (perhaps) the breaking off or the breaking away happens; but perhaps not. Emotion is vested in both breaking and not breaking, in both the unity and the assertion of uniqueness. Yet uniqueness is not asserted through and in the clash of individuals. Friction is not a clash; it is not yet a (the) clash. Those famous balls (or spheres) of Cusanus are still moving within the same circle, although their motion is different subsequent to their original "impetus." It is only outside the circle that there is no unity, no trust, and also no discomfort; just alienness, indifference, hostility, and hatred. But there is no friction.

Friction and healing can be considered together. Healing is certainly a metaphor of biological provenance that has acquired metaphoric prominence since time immemorial. In combining the metaphors *friction* and *healing*, one is able to reclaim certain aspects of the mystical appeal of the latter. In *The Mystical Foundation of Law*, Derrida reclaims the metaphor *force* for such a mystical use.

To avoid misunderstanding, thinking of friction-as-metaphor and healing-as-metaphor together is one thing, but whether one or another actual friction can be healed or ends with breaking up or breaking off is an entirely different matter. I discuss imagination. Mythical imagination revolves around the winning and losing of the battle of survival and power, whereas mystical imagination revolves around the One as the Individual in its constant breaking off from One as Unity, and around the healing of frictions. The poor, melancholy metaphor of *friction* is utopian because it accepts the double bind.

PART EIGHT

Into the Future

17

Frictionless Forecasting Is A Fiction

Hubert L. Dreyfus and Stuart E. Dreyfus

Observers of the political and economic scene note that almost all decisions involve incremental changes from the status quo.[1] Slightly mitigating the ills we have has always seemed preferable to flying to others that we know not of. It now appears, however, that advanced information technology and improved theoretical understanding of social and economic phenomena may have put us on the verge of a breakthrough, and the possibility exists of flying directly to radically new solutions that we can predict sufficiently well, not only to avoid ills but, discontinuously, to enhance the good. Arguments for this brave new frictionless society go roughly as follows:

1. With the increase in specialization, experts are available today who are capable of short-term forecasts about almost all aspects of social and economic behavior.
2. Concurrently, it is becoming increasingly difficult for any one person to integrate all of this understanding.
3. Computer technology now allows us to incorporate this immense body of short-term forecast knowledge into programs that, through simulation, can predict the long-term future impacts of contemplated decisions. These impacts are often surprising and counterintuitive, accounting for the fear, in the past, of anything but small changes.[2]
4. Once we can predict the impact of globally discontinuous decisions, debate can appropriately center on which effects are preferable rather than on what the impacts will be.

The benefits of successfully implementing this conversion to frictionless decisionmaking are obvious. In what follows we shall, regrettably, develop reasons for doubting that such a program can be implemented successfully, currently or in the foreseeable future, and we shall identify risks, beyond the obvious one of inferior decisionmaking, inherent in moving toward such an unattainable goal. If incremental change is indeed to remain man's fate, we shall ask how the emerging power of information technology can be harnessed in such a way as to improve at least incrementally the prospects for better incremental decisionmaking.

Computer models of physical phenomena, built up on the basis of contributions by many scientists specializing in minute details of the physical domain, have been the basis of most recent technological developments, including the astounding evolution of the computer industry that is reshaping our world. The hope that the same process will succeed in the socioeconomic sphere spurs the quest for frictionless predictive socioeconomic modeling. But is it reasonable to extrapolate from the science of physics to the socioeconomic domain? Unfortunately, due to the complexity, uncertainty, and nonreplicability of real-world socioeconomic situations, computer models of such phenomena can never be tested and refined the way models of physical phenomena are. Even if a model should prove reliable for a time in its short-term predictions, there is no way of knowing whether some overlooked aspects of the situation are fortuitously remaining unchanged so that an unanticipated change in one of them may at any moment render the model invalid. Of course, at the other extreme, human intuitive experts, arriving at their own, generally incremental, predictions and recommendations in an experience-based but otherwise inexplicable way, are likewise certainly fallible.

Due to the immense and open-ended variety of possible real-world situations, the impossibility of artificially constructing situations for testing purposes, the large number of both expert opinions and computer-model predictions spanning almost all possibilities, and the comparatively small amount of case data on which to base any comparisons, disagreement over whether models or experts are more reliable can never be resolved, as conflicting approaches in physics often are, by simple comparison of performance. If any light is to be shed on the controversy, it must come from examining and questioning the reasons for supporting each camp. Unfortunately, arguments for believing or distrusting socioeconomic models are seldom presented or discussed. Both sides take their views concerning the worth of models versus expert opinion to be self-evident. We shall argue below that implicitly held beliefs about the nature and value of rationality and the nature of expertise largely account for the conflicting conclusions.

The Nature of Intuitive Expertise

Our aim is to distinguish four views of the nature of skillful human coping and predicting, describing rather briefly three of them—those held by various schools of thought—while articulating our own position in greater detail. We then present our reasons for doubting the three competing views. Certain observations concerning skillful behavior are, we believe, acceptable to all four views, so we shall start by stating them. Clearly, beginners do not perform well, and with training and experience performance improves. In animals considerably lower on the evolutionary scale than man, it is fairly certain that behavior that is not innate is the result of trial-and-error experience, with synaptic modifications in the brain reinforcing successful behaviors while inhibiting others. Indeed, these synaptic changes cannot be even approximately described at some high level of abstraction such as belief, goal, or mental model of the domain. Matters become more complicated and controversial, however, when they concern skilled human behavior. Our trained-in and imitative social comportment and the physical movement of skilled laborers such as carpenters are probably best seen as analogous with, but of course much subtler than, lower animal coping behavior. We need have no theory or mental model of our social or largely physical vocational skills in order to learn through trial-and-error and through instructional example to act acceptably and even skillfully when, for example, involved in carrying on a conversation or hammering a nail. Involvement in *real* situations seems crucial to this effortless and usually successful behavior, for if we are given a (necessarily incomplete) verbal description of a conversational or hammering situation, what, after conscious deliberation, we say we would do or would expect to have happen is unreliable and even varies dramatically with differently worded descriptions of the same situation.

All accounts of skilled coping would also probably agree, on the other hand, that our detached "problem-solving" comportment, when we are beginners in a new and largely cognitive domain or when we are faced with entirely novel situations in domains in which we have already acquired skills, seems to be at least approximately describable at the abstract level as "reasoning about the situation based on a theory of the domain." What novices decide they would do or would expect to transpire when a situation is described to them in terms of what has been taught are the salient features of such a situation, is usually what they would actually do or predict in such a situation.

Most of the vocational activities of social planners and other so-called knowledge workers fall between these extremes of synaptically associated response and detached problem solving, and it is here that the frictionless potential of modeling is most seductive but, simultaneously, most controversial. Planners, as well as businesspeople, surgeons, teachers, and so on, cope with, or make short-term predictions about, plausible incremental changes fairly effortlessly and, much of the time, successfully in situations that are hardly novel, yet not identical with ones previously experienced. They do so, particularly if time is short, with no conscious awareness of problem solving, and even when time permits they deliberate more about the relevance of prior experience and the possibility and risks of overlooked alternative perspectives or available facts than about rules and principles underlying their skill. Observers of the social-planning domain, in particular, would also probably agree that experts are not especially reliable when it comes to long-term predictions concerning the implications of even incremental changes. The hope for computer modeling as the means of attaining a more frictionless society is founded on formalizing this short-term expertise and using it as the basis for computing accurate long-term, possibly counterintuitive, predictions. The successful skilled short-term behavior that occurs in familiar sorts of real-world situations, therefore, is what interests us here. It can be "explained" in at least four different ways.

The position implicitly held by most workers in the field of artificial intelligence, especially those developing expert systems, might be called *symbolic/rationalistic*. The skilled performer is assumed to have acquired from early training, and modified by subsequent experience, a theory of the skill domain. This theory, when used for *prediction*, takes the form of rules for what elements of the situation are relevant in a situation and for how these elements will change with time as a function of decisions taken. Skilled *performance* is the result of applying further rules to determine a goal given the relevant elements and then seeking decisions that transform the current situation into one that satisfies the goal. This picture of skilled performance differs from the generally accepted explanation of beginner behavior in two ways: The beginner chooses not goals but merely actions based on the elements of the situation, and the beginner uses less sophisticated rules, generally those taught rather than those learned from experience. Most advocates of the symbolic/rationalistic position would agree that, while the beginner's procedures are conscious, the more experienced and skilled performer is not consciously aware of the rules being applied. The art of deducing these (assumed) rules from observation and interrogation is called "knowledge engineering." Successful knowledge engineering, combined with a large data base of facts, is seen as the key to programmed expertise.

At the opposite extreme from the symbolic/rationalistic view lies the *subsymbolic/associationistic* one. This perspective sees expertise as purely the result of experience, and its explanation of skilled behavior is based on speculative neuroscience. In this view, a performer, immersed in a real-world situation, has experienced a time-series of stimulate that has caused the brain to exhibit a certain pattern of activity. If the output neurons of the brain then produce an action or thought that turns out well, synaptic modifications occur that reinforce the same result in future similar brain-states. Negative reinforcement (inhibition) follows unsuccessful behavior. From this perspective, the process of mapping received stimulate onto produced outputs has no interpretation in terms of learned rules or chosen goals. That is, a skilled performer has acquired through reinforcement and inhibition a pattern of synaptic connectivity that generally produces successful output, but does so without reference to any rule-based, theory-driven procedure. When new situations are being predicted from current ones, feedback concerning the accuracy of the prediction is immediate. But when long-range effects are being predicted, or decisions are being evaluated, feedback concerning success or failure of a behavior occurs only after a *series* of stimulate-produced outputs. This theory, then, requires a solution for what is called the "credit assignment" problem: Which associated outputs should be credited or blamed for the final result of the series of behaviors? Progress has recently been made in explaining how the brain can solve this problem automatically, using what is called reinforcement learning with an adaptive critic, so advocates of the subsymbolic/associationistic view are not overly concerned about this possibly troublesome issue.[3]

We come now to two more plausible explanations of expertise that lie between these extremes. Saving our own view until last, let us examine the *case-based* explanation.[4] Advocates of this position, such as Herbert Simon, argue that concrete experiences themselves, rather than rules and theories derived from these experiences such as postulated by the symbolic/rationalistic camp, seem to guide behavior. They explain the role of experience, first, by observing that experts have seen up to perhaps 50,000 situations in their domains of expertise and, then, by speculating that these situations, behaviors, and results are separately and explicitly stored in the expert's memory. (Note that the subsymbolic/associationistic explanation hypothesizes, not separate storage of experiences, but synaptic modification based on these experiences that strengthen successful response to situations sufficiently similar to the experienced situations.) Given this vast repertoire of stored memories, incoming elements in a situation are compared to the elements characterizing each of these experiences, the most similar experience or experiences are accessed, and the behaviors and results for these cases are used to guide behavior in the current situation. Implementation of this approach requires a basis for describing situations in terms of their elements, a rule for determining similarity of situations, and some sort of logical procedure for synthesizing the most similar experiences to produce expert behavior.

Our view of the acquisition of skill by knowledge-workers initially relying on instruction is multistaged. It is based on observations of the skill-acquisition process of nurses, pilots, and chess players, and on introspection. We believe that the process of skill acquisition undergoes marked qualitative changes as it evolves. Following is a brief description of five such stages that we have observed, in which short-range prediction is used to illustrate the stages.[5]

Stage 1: The Novice

Normally, the process begins with an instructor decomposing the task environment into context-free features that the beginner can recognize without benefit of experience. The beginner is then given rules for determining actions or predictions based on these features. The beginning student wants to do well but, lacking any coherent sense of the overall task, judges performance by how well he or she follows the learned rules. After acquiring more than just a few rules, great concentration is required during the application of the rules. A student of economics, training to forecast interest rates, might learn that certain economic indices and indicators are relevant to the situation and be given a formula for using these facts to predict the direction and size of interest rate changes during a certain future period. Novice behavior, most cognitive psychologists would agree, is adequately described by the symbolic/rationalistic conception of behavior.

Stage 2: The Advanced Beginner

As the novice gains experience by observing real situations, he or she begins to note, or an instructor points out, perspicuous examples of meaningful additional components of the situation. After seeing a sufficient number of examples, the student learns to recognize these components. Instructional maxims can now refer to the new *situational aspects* recognized only on the basis of experience, as well as to the objectively defined *nonsituational features* recognizable by the beginner. The advanced beginner confronts the environment, seeks out features and aspects, and determines actions or predictions by applying learned maxims. He or she shares the novice's minimal concern with quality of performance, instead focusing on quality of rule following. The advanced beginner's performance, though improved, remains slow, uncoordinated, and laborious.

For example, the advanced beginner student of economic prediction might become aware that consumer optimism, among other factors, affects buying behavior and, indirectly through demand for money, interest rate behavior. After studying or experiencing examples of various degrees of consumer optimism, the student learns to assess optimism in the sorts of situations observed and to incorporate this factor into his or her predictive rules.

Recognition, based on experience, of situational aspects can be seen either as the acquisition of unconscious necessary and sufficient conditions (the symbolic/rationalistic explanation) or as the product of synaptic changes resulting from exemplars with no intervention by what might be called symbolic/rationalistic rules (the subsymbolic/associationistic explanation). We give our reasons for preferring the latter view below.

Stage 3: Competence

With increasing experience, the number of features and aspects to be taken into account becomes overwhelming. To cope with this information explosion, the performer learns, or is taught, to adopt a hierarchical view of predicting or decisionmaking. By first choosing a perspective that organizes the situation and determines which elements should be treated as salient and then examining the small set of elements deemed salient given the perspective, the performer can simplify and improve performance.

Choosing a perspective is no simple matter for the competent performer. It is not a safe, objective procedure, like the feature recognition of the novice. Nobody gives the performer useful prescriptions for reasoning out the appropriate perspective, since there are too many possible situations and, as we shall see later, more proficient performers do not use articulable formulas and reasoning to achieve this end. The performer has to make up various rules, which are then accepted or discarded depending upon how they turn out. This is frustrating, however, since each rule works on some occasions and fails on others, and no set of objective features and situational aspects correlates strongly with these successes and failures. Although the advanced beginner can get along without recognizing and using a particular situational aspect until a sufficient number of examples makes recognition easy and sure, competent performance *requires* choosing a perspective; hence the choice, though risky, is unavoidable. Furthermore, the choice of perspective crucially affects behavior in a way that one particular aspect rarely does.

This combination of necessity and uncertainty introduces an important new type of relationship between the performer and the environment. The novice and advanced beginner applying rules and maxims feel little or no responsibility for the outcome of their endeavor. If they have made no mistakes, they can view an unfortunate outcome as the result of inadequately specified elements or rules. The competent performer, on the other hand, after wrestling with the question of a choice of perspective, feels responsible for, and thus emotionally involved in, the result of the choice. An outcome that is successful is deeply satisfying and, if he or she rises above competence, directly affects future behavior. Disasters, likewise, leave an indelible mark.

For example, a competent economic forecaster might consciously decide, in a given situation, whether consumer confidence or the world political situation or both, along with other elements of the situation, have been crucial in impacting interest rate behavior and then arrive at a prediction by combining the effects of crucial considerations while ignoring or downplaying what were deemed to be less critical influences. In this and all such cases, the competent performer, in a detached, rational manner, first decides on a perspective, then assesses those elements that are crucial with respect to that perspective, applies a learned rule to arrive at a conclusion, and enjoys or suffers an emotionally involved experience of the outcome of events.

We believe that, except when a subsymbolic/associationistic process is providing situational aspect recognition, the competent performer's behavior is correctly described by the symbolic/rationalistic model.

Stage 4: Proficiency

Considerable experience at the level of competency sets the stage for yet further skill enhancement. Having experienced many situations and chosen perspectives in each, and having obtained vivid, involved demonstrations of the adequacy or inadequacy of the choice, the proficient performer begins to spontaneously associate an appropriate perspective with situations when they are similar to those already experienced. Involved in the world of the skill, the performer "notices," or "is struck by," a certain perspective. Due to association, the spell of involvement is not broken by detached, conscious choosing.

Our speculative explanation of this remarkable new event is that the synaptic brain connections that supported the rule-based chain of inferences of the competent performer have been modified through experience so that the brain mapping of situation onto perspective is no longer conscious and decomposable into articulable rules. It is simply an input-output mapping process that, due to its refinement through experience, generally works. At this point the symbolic/rationalistic instantiation in the brain of the symbolic model erodes until it is unrecognizable and only a subsymbolic/associationistic model remains. Note, however, that the synaptic implementation of rule-like behavior was the starting point for this transformation, so early theory-based training as well as subsequent experience are responsible for whatever perspective is taken by the proficient performer.

Since there are far fewer ways of viewing matters than predictions or decisions available, after associating a perspective with a situation the proficient performer will still have to think about what to do or predict. During this conscious process, the elements that present themselves as salient are assessed and combined by rule to produce decisions or predictions. It is here that the spell of involvement in the world of the activity will thus temporarily be broken.

Seeing what is important in the current economic environment, without being able to rationalize the ability, the proficient forecaster decides how these considerations will impact future interest rates.

Stage 5: Expertise

As we have seen, the proficient performer, immersed in the world of his or her skillful activity, *sees* what is going on, but *decides* what to do or what this implies for the future. For the expert, not only this situational understanding springs effortlessly to mind but also associated appropriate actions or predictions. The expert performer, except of course during moments of breakdown, understands, acts, and learns from results without any conscious awareness of the process. What transparently, to the expert forecaster, is going to happen is what he or she foresees. Thus the expert is acting fully in accord with the subsymbolic/associationistic paradigm. It should not be forgotten, however, that he or she got there from a background of rules and theory learned at earlier stages of development.

Obviously, in a given domain some experts will be able to discriminate more situations than will others. The boundary between circumstances that are responded to in one way and those responded to in another differs for various experts and is crucial to the quality of performance. Also, the responses that some experts associate with these discriminable situations will be more appropriate than those of other experts. The best of experts combine finer discriminations with well-placed boundaries and accurately tuned responses. These masterful experts also deliberate about their intuitive responses in the manner briefly described earlier in this section and elaborated elsewhere,[6] thereby possibly improving them further. And, of course, in areas such as economics or social planning, even the best of experts are all too often wrong, but, we believe, perform on average at a higher level than those at earlier stages of skill development. (This is demonstrable in areas we have studied such as chess playing, nursing, and piloting.)

The expert economic forecaster, gathering whatever information is available, will simply conclude that interest rates are likely to follow a certain trend and will recommend actions accordingly. Although the expert may try to explain his or her conclusions in terms of rules and principles, if these same rules are applied to a different situation encountered later they are unlikely to generate assessments that agree with the expert's assessment of the new situation. (This conclusion has been clearly established in the area of chess, where computer programs implementing the rules and principles of plausible-move generation provided by grandmasters performed at nowhere near grandmaster level; indeed, it was only after this approach was abandoned in favor of brute-force enumeration of all legal moves that high-level computer chess was achieved.)

We promised earlier to give reasons for accepting something resembling our skill-acquisition model over the three alternatives. The main reason, we believe, is that it fits our observed and introspected experiences, and readers should certainly ask if it seems consistent with their own. To us, the symbolic/rationalistic description seems only part of the story, fitting well with the behavior of laboratory subjects being observed while learning a new skill, but not with the behavior of an expert at work. An expert's responses are too fast and effortless to be the product of the chaining of inference rules. Furthermore, animals are experts in the domains relevant to their survival and comfort, presumably without recourse to symbolic models and inference rules; likewise, infants and young children attain abilities critical to their comfort prior to reaching the age of reason. Finally, the repeated failures of the field of artificial intelligence, using symbol manipulating and rule-based programs, to produce truly expert behavior (even in the misnamed subfield called expert systems), even though the speed and rule-following ability of the computer are far superior to those of the brain, seems to us to be confirming evidence for the inadequacy of the symbolic/rationalistic hypothesis.

The subsymbolic/associationistic position suffers from what is called the "generalization problem." If only synaptic changes due to experience mattered, one would expect all experts with shared experiences to more or less agree (assuming they were all born with similarly arranged brains) or else to all have quite different opinions (assuming their initial neuronal configurations were quite different). Yet it is a fact that there are schools of experts who see things similarly within each school but quite differently across schools. It seems to us that the most plausible explanation of this phenomenon is that the shared experiences have modified the brain-embodiments of quite different school-dependent domain theories, and that to ignore the role of early theoretical training is to fail to do justice to the phenomenon. Thus we are led to the conclusion, not necessarily shared by the subsymbolic/associationistic perspective, that early trained-in points of view drastically affect the way in which experience will ultimately produce behavior.

The case-based view that separately remembered experiences are accessed and combined to produce behavior fails to explain the ease and speed of expert performance. It is also inconsistent with the way memory and performance gradually degrades with slowly progressing brain damage. This perspective, as well as the subsymbolic/associationistic one, is currently being suggested as an alternative approach to artificial intelligence in view of the failure of the symbolic/rationalistic school. Reasons for doubting either's success will be presented in the next section.

Capturing Expertise in Computer Programs

If a computer program could, with high quality, predict the short-term effects of contemplated decisions in a large variety of different situations, it could extrapolate this knowledge well into the future and thereby prove to be an invaluable aid in choosing long-range social and economic policies. (This is analogous to the fact that knowledge of the derivative of a function for any argument value allows one to determine the function computationally, i.e., to integrate a differential equation numerically.) Therefore, to decide whether experts and computers together can accurately turn short-term predictive ability into accurate long-range forecasts, we need to ask if the ability of the expert in these matters is sufficient and if it can be transferred to the computer. The answer depends to some degree on our view of expertise, but to a surprising extent the answer is in the negative no matter which view we adopt.

As we have seen, the symbolic/rationalistic model of skill views inference rules learned from experience as necessary and sufficient for quality performance. These rules say that if such and such is the case, so and so will happen or perhaps that this and that fact is crucial to the situation. If such rules exist in the mind of the expert and if knowledge engineers can somehow extract them, they can, of course, be programmed into the computer. The program then should do as well as the expert in situations where the rules apply. The question is whether these rules apply in situations quite different from those that led the expert to acquire them. Suppose that, based on experience with incremental changes in tax structure, an expert can accurately say that if the tax rate is raised 2 percent in the current situation certain changes will occur in consumer behavior that will cause the interest rate to fall 1 percent. If a doubling of the tax rate is contemplated, this rule in itself will not help the computer. Only if the rule is restated to read "An x percent change in tax rate will produce a $-.5x$ percent change in interest (or some other formula whereby a 2 percent increase in tax rate yields a 1 percent decrease in interest rate) would it be applicable to the contemplated doubling of tax rate". But why should such a global application of a locally learned rule be believable? Certainly a degradation in the accuracy of the forecast would accompany such an extrapolation of knowledge, and this degradation is the friction accompanying the procedure. Even in the short term, the rules and formulas of the hypothesized symbolic/rationalistic expert can at best be expected to be accurate only for the sort of incrementally changed situations that the expert has experienced.

Furthermore, long-term forecasting of even incremental changes can be trustworthy only if the theory of the domain of expertise that members of the symbolic/rationalistic school believe the expert knows is applicable over a wide range of values of elements describing a situation. But a socioeconomic expert has experienced only a small subset of possible situations and cannot, in the manner of the physicist, construct and experiment with others to determine the extent of applicability. Hence long-term forecasts of even incremental changes in policy, or of no changes at all, are suspect if, in the long run, the situation changes significantly from the status quo. One has to believe that an expert, given a description of a situation quite different from any experienced, can miraculously apply rules to make expert predictions of short-term behavior if one is to have faith in long-range computer forecasts in general. The faith in computerized expert-quality short-term forecasting of the impacts of incremental changes is more plausible, although we should recognize that it too gains its plausibility from the symbolic/rationalistic view of expertise. If, as agreed upon by the other three views of skill, experts do not use articulable rules to generate their short-term predictions concerning incremental changes, the knowledge engineer is attempting the impossible when probing for such rules.

Adherents of the subsymbolic/associationistic school would, of course, claim exactly this. From their perspective, rule-based predictions with even short-term validity merely for incremental changes are suspect, and the dream of rule-based frictionless understanding of long-term implications of radically changed policies is baseless. Engineers and computer scientists in this school might, however, believe that through the construction of artificial devices operating much as our biological neural system does, with learning from experience realized through modification of strengths of connections between artificial neurons, automated expert-quality short-term predictions concerning situations similar to those used for training these devices could be achieved. In fact, such attempts are currently being made and the field of artificial neural systems is booming.

What are the prospects for success of this endeavor? To succeed, the field needs to solve what is called the "generalization problem." The human brain probably contains well over a billion modifiable synapses, and to have comparable abilities an artificial neural net might well need to include millions of such adjustable parameters. Even the learning of tens of thousands of appropriate outputs (predictions), given the sheer number of potential inputs (situations), fails to determine uniquely or with even approximate uniqueness these parameters, which, of course, determine the response to other inputs. Hence, given the same training cases, the details of an artificial neural net such as the initial configuration and synaptic connections before training begins, the representation in terms of neuronal activity of the input and output, the parameter modification procedure used during learning, and so on, can yield final trained nets agreeing on the training cases but differing widely in their responses to new situations. If no two nets generalize from the same experience similarly, there is little reason to trust any particular net's responses to new cases. It seems to us that what is lacking in current neural net devices is, perhaps among other things, the ability to learn the conceptual rules that beginners and competent performers acquire prior to synaptic modifications based on experience. Until this ability is achieved, and it is a very difficult research problem, artificial neural nets are unlikely to produce predictions of consistently expert quality.

Suppose that, as the case-based school proposes, experts store separate memories of tens of thousands of situations and associated successful actions or correct predictions and use these to guide behavior in new situations. Suppose further that one can represent these cases in terms of features and their values and, after representing the current situation in like manner, prior situations can be identified that, by some measure, are similar to it. Then, if the responses (i.e., actions or predictions) of these most similar cases can somehow be combined to produce a response, stored experience can be used to generate expert output. The above program leaves unanswered the questions of what measure of similarity to use, how to combine possibly contradictory responses associated with similar cases, and what criterion to use if several different responses are to be compared and one chosen. Only if experts do indeed respond to situations in the manner assumed by the case-based school is there reason to believe that answers exist to the above questions that will produce expert-level performance. Even then, since experts are unaware of using such a memory-based procedure, the modeler would have to guess at answers or else plumb the expert's unconscious. These practical roadblocks, if this model of expertise is to be believed, compounded by the reasons for distrusting this view of expertise given earlier, lead us to reject this endeavor as misguided. And, of course, even if this type of modeling were to work, it could only be expected to produce good short-term predictions for situations similar to those experienced, thereby falling far short of the goal of long-term predictions concerning radically differing policies envisioned by those hoping to overcome forecasting friction.

What *Can* Computers Do?

Assuming, with us, that long-range prediction of socioeconomic phenomena, with respect both to current policies and to radically different ones, is beyond the pale of both computer model and expert, that even short-term predictions by computers or experts of the impact of policies differing greatly from those previously pursued are not to be trusted, and that short-term predictions concerning current policies or those differing incrementally from them are better left to experts than to computer models, one might naturally ask: What then do computers have to offer in the policy-analysis domain?

Computers are being used and should be used as data providers, manipulators, and displayers. We live in an age of information explosion. But if expertise is based on intuitive synapse-based discrimination and association and not on data-driven rule-based computation, of what use are all these data? The answer, we believe, is that a current situation can now be known in far greater detail than ever before, and that this enhanced situational description potentially allows the human expert to discriminate more finely. We should now be asking: What information, in what form, should experts be given in order to improve their expertise in a particular domain?

It might seem that this is an empirical question. Just give experts of similar abilities and experience differing pieces of information in differing form and see who, on average, does the best. But paradoxically, our model of expertise implies that any expert who attempted to incorporate in his or her predictions the information provided, no matter what kind of information and which form it takes, would suffer a degradation in performance. The only way to incorporate new data, which the brain has never before received, is to forsake the synapse-based neural mapping process that we speculate produces the expert's skill in favor of a reasoning process that is at best competent. Using new information is itself a skill that only considerable experience with that information can produce. The real research issue is this: Assuming that experts of comparable ability and experience are made privy to different new bodies of information displayed in some way, and that they are each given a long period during which to acquire the skill of using this information in determining their intuitive responses, what information, displayed in what form, will produce, on average, the highest quality expertise?

To make this matter precise, imagine a group of skilled air traffic controllers, all experts at using the standard displays and facts to produce actions. Several new devices for displaying further data and making accessible further information are proposed and the one most helpful to the skill is to be adopted. If the current equipment of each controller is replaced by some different proposed device, the performance of each will most certainly initially degrade, probably dramatically, as the controllers competently reason out their actions based on their new data and displays. Eventually, some, probably all, of the controllers will become intuitive experts at using their new equipment and some, perhaps all, will produce actions of higher quality than when the experiment was begun. In this way, and, we fear, only in this way can nonincremental changes be evaluated. Since it is not feasible to relieve a number of expert controllers of their duties to carry out this experiment, the frictionless introduction of major computer-based enhancements of work conditions may not be possible. Perhaps even when appropriate use is made of the information-handling abilities of the computer, only incremental changes are feasible.

Dangers in Trying to Avoid Friction

Developing a set of rules and principles that produce expert-quality performance in an entire domain of skill is, in our view, synonymous with overcoming friction in that domain. As discussed, should such a set of rules be found, perhaps based on the knowledge of a large body of experts, sudden and dramatic improvements could be realized. We have argued that no such rules exist, however, and that, therefore, the quest is a chimera. Worse than that, simply *believing* the goal to be achievable is menacing to society.

According to our model of skill acquisition, proficient and, finally, expert performers, involved deeply in their skill and experiencing the outcomes of various events with emotion, naturally become more skilled due to synaptic reinforcement and inhibition. This is a normal process, certainly true of animals and of everyone's everyday behavior. But it would not be true of competent knowledge workers who remained detached observers of their skill and who saw each experience as grounds for adjusting their rules and procedures. They would be inhibited by their view of their skill from ever trusting their intuitions, from becoming involved, and from progressing to higher levels. In the past, unaware and unconcerned about what underlay their emerging skill, few learners were inclined toward this fate. But now, in domains where the expert system, with its inference rules and vast database are held up as the model of expertise, the danger suddenly exists that learners will find their role model in the expert system, aspiring at best to acquire more subtle rules and principles through experience in order to outperform the system. For example, if doctors trained to competence by emulating rule-based expert systems, internalizing as their goal the system's approach to decisionmaking, they would never become experts. The expert system may not merely fail to produce genuinely expert behavior, but, even more frightening, human intuitive expertise itself will become an endangered species if detached inference making becomes our model of expertise. Perversely, the quest for artificial intelligence through rule-based behavior holds the potential for undermining our true natural intelligence. The attempt to overcome friction, then, may be creating new frictional forces that impede what progress we could otherwise incrementally achieve.

PART NINE

Metaphor Transferred

18

Rediscovering Friction: Not All That Is Solid Melts into Air

Helga Nowotny

From Modernism to Postmodernism

Modernism ceased to be fashionable some time ago. To live a life of paradox and contradiction, to be moved at once by a will to change and to be horrified at the same time by the prospect of disorientation and of life falling apart, has ceased to be a meaningful way of how today's men and women experience the dramatic changes around themselves and in themselves. Yet, a curious split of consciousness seems to occur. In one part of Europe, a kind of premodernism has apparently emerged, characterized by the reappearance of primordial ties of real or imagined ethnicity and mutually exclusive belongingness, each group ready to deny to others what it claims for itself. In another part of Europe, the supposedly postmodern turn is undergoing equally troubling, though far less painful, convolutions to reach a more encompassing stage of integration, this time primarily in the name of a greater economic unity to which other forms of unification might follow.

One half of a suddenly rearranged and enlarged Europe is actively engaged in deconstructing space and time after having been occupied for so long by central powers that no longer exist, while the other half is deconstructing space and time as part of a welcomed process of internationalization and further technological modernization. In the Eastern part, territorial boundaries and neighborhoods are violently rejected and resisted in the name of fictitious or real historical events. The past, including memories of it, pregnant with meanings that militate from their inception against compromising solutions, has become unfrozen after decades of centralized terror, great and small. Ironically, in the territories of the former Soviet Union, with the dissolution of every law except that of bureaucratic inertia and the nonrule of those acting outside the law, nearly the only identities to cling to are the ethnic ones assigned to the empire's subjects in view of facilitating their deliberate reallocation to territories other than those they initially lived in. Now, with the former nightmare having come to an end, a plethora of revengeful nationalisms has spread, claiming territorial rights for minorities where majorities can no longer be clearly defined. Yet, the will to change is limited to one's own ethnic group and limited to numerous minorities living amid other minorities and making up majorities, while a new horror of disorientation and disintegration spreads. Modernism has apparently become involuted in premodern attachments in the name of the primordial ties in which it continues its work of dissolution.

In the other part of Europe, the affluent, pacified, highly industrialized Western part, modernism has also ceased to be meaningful. It has moved into a stage of transformation that, for lack of a better term, is commonly referred to as postmodernism. Here it is not so much space, defined as territory, and time, defined as history by those who occupy it or raise claims to territorial occupation, that provide the coordinates for regression or progression. Rather, the deconstruction of space and time proceeds in line with the expansion and compression of space-time as enabled and effected by the means of modern, or rather postmodern, technology and economic competitiveness. Territorial claims as the primordial mode of attachment and identity-giving recede in significance wherever and whenever satellite connections are available, with remote sensing in operation and infra-red detection distinct possibilities. What are being divided and shared are not territories in which mostly decaying and outdated industries are located but, rather, markets and the purchasing power that makes them attractive.

The deconstruction of space and time in the West is turned outward, as well as inward; it aims for control through speed and networking, market shares and deregulated competitiveness, on an international as well as national and local level. In both a lateral and vertical fashion, global interconnections and local accessibility are to be attained. All those who can afford it are to be integrated or are induced to strive for integration. Modernism, as the will to change and the horror induced by the unforeseeable results that change may bring, defined by Marshall Berman as the voice that is "ironic and contradictory," resonating at once "with self-delight and self-doubt," that "denounces modern life in the name of values that modernity itself has created, hoping—often against hope— that the modernities of tomorrow . . . will heal the wounds that wreck the modern men and women of today" has been surpassed here as well (Berman, 1982). It has been replaced by the so-called postmodern condition, a curious aspect of the modern self, which has become dissolved into a fragmentary mode of existence whereby self-doubt has extended to the core of the self, erasing any self-delight. Hope has vanished and so has the belief in the future, any future. Instead of an affirmative vision, a feeling of the futility of it all has spread—with pockets of hedonistic survivors striving to cling to their postmodern existence, embedded in collectively unsurpassed affluence and the deeper anxieties that come with it.

What we are witnessing today is indeed a dramatic stage of transition, and not only to market economies where there were command economies before. The social sciences in particular are challenged to observe, analyze, and interpret. They are challenged to respond to the unprecedented experiences that are historically novel while these changes occur under the eyes of the observer. What is denied to them is the benefit of hindsight. For once, and in this they resemble their predecessors witnessing the profound turbulence of nineteenth-century modernism, they cannot claim the benefit of hindsight, the scholarly security and distance, that comes with time having lapsed.They are continuously confronted with the new. The scholarly vantage point of historical distance is shrinking, while the temptation to compete with journalistic accounts, written from one day to the next, has perhaps never been so great. Before 1989 at least, the general contours of the changes appeared reasonably clear. Modernity had surpassed itself and was being dissolved in a multitude of postmodern epiphenomena. Whereas some would claim that in the process modernity with its Enlightenment roots has also been defeated and that progress of any kind becomes impossible, others would more cautiously assert that modernism still exists side by side with its postmodern mutants and variants. The contours of the major shifts are agreed upon, while the interpretation naturally differs. What has been shed is the belief in progress, an unending vision of human betterment that had been intimately linked with the unfolding instrumental effectiveness of scientific and technological knowledge. What has been lost is any central perspective that can claim superior knowledge or command control: In its stead, a multitude of particularistic perspectives claim rights to both.

Behind such shifts in beliefs are other, more substantial changes that extend to the foundational transactions of the modern welfare state's arrangements. The State itself, as a powerful centralized actor, has come under attack from within; it is encountering its own limits of action and outreach. It is also challenged outside its own boundaries to meet increasing demands of internationalization and cross-boundary, diffusely lateral interlinking of multiple networks, private and public alike. The production system has dramatically lost its centralized and hierarchical features. Not only is it no longer possible to plan and centralize the many functions that a modern or, rather, postmodern enterprise has to carry out, but the industrial process has also moved from working with bulk material to material designed to serve certain specified functions, possessing pre-designed and desired characteristics. Slowly, but surely, the control of waste is moving from an inefficient output control to a somewhat more efficient input control. Mass production, although it still accounts for a sizeable proportion of all production processes and products, is becoming complemented by production according to design. The process of post-Fordist industrial transformation, as it is called, in which the logic of standarized mass production and mass consumption, based on the mass worker, is giving way to various forms of flexibilization and deroutinization. The manufacturing system is increasingly built around so-called nonmaterial inputs and specialized technical and professional services (Esping-Andersen, 1992). Flexibility, rather than centralized control, has become the key word for management styles, along with new fashionable buzzwords in praise of unpredictability and adaptation to unforeseen circumstances. Chaos is not only a mathematical theory but has become a cherished belief system for management—a positive form of adaptation to what cannot be controlled, a seemingly self-explanatory worldview that, by providing the necessary minimum dose of certainty still attached to a scientific worldview, promises to alleviate anxieties stemming from a rapidly changing environment.

The altered relationship between State and the Economy, their relative positions in absolute and relational terms, cannot but affect the lifestyles of citizens. Ordinary men and women in their roles as producers and consumers, as voters and environmentally conscious habitants of Planet Earth, are confronted with equally altered lifestyles from which they are told they can choose, while the previous regularity and predictability of their working-life careers, class-bound as they were, give way to new and much more erratic patterns, marking many more interruptions on the road from youth to retirement. Whereas the Fordist industrial order was built around a very particular gendered division of labor, centered around the male breadwinner's full-time industrial wage employment, the post-Fordist, postindustrial stratification system may see a convergence of female and male life-cycle profiles, characterized by a spillover of the hitherto more precarious, temporary employment arrangements of women to their male counterparts (Esping-Andersen, 1992; Schuller, 1988). And while the clear-cut stratification profiles of the Fordist-industrial stratification system with its relatively straightforward and predictable chances for upward and downward social mobility is giving way to another hierarchy, the dimensions have multiplied, and its occupancy certainly has become more transitory and volatile (Reich, 1991).

On the Borrowing of Metaphors:
Transfers Between the Two Cultures

It is against such an impressionistic background of fluidity and change from one central perspective to multiple ones, all rooted in their own uncertainties and in radically altered subjective experiences, and against the rapid loss of meaning of older social science concepts, such as social class, that the rediscovery of a concept like friction has to be evaluated. It is a concept that—in a self-applicable way—intrinsically stems the tide of flexible postmodernity, in which we are made to believe that everything becomes fiction, self-reference, signifier, and ephemerality, and in which immateriality reigns supreme and homelessness, chance, fictitious capital, and eclecticism are sufficient to replace public housing, purpose, production capital, and authority. (For the complete list of replacements of opposed tendencies, see Harvey, 1990, pp. 440–441.) Without posing any moral or philosophical preconditions, interaction is a concept that is central to the natural world and to social life. In the macroscopic world, frictionless interaction is a rare exception. Nowhere in the universe, neither in the cosmos nor in our own social world, can anything happen without interaction—often, interaction entailing friction. In many cases, self-organization, too, owes its existence to precisely this. There can hardly be a social world, however anomic it may appear, without interaction and friction. Interaction occurs even in the breakdown of social relations, whether in the climax of physical violence directed against other human beings—war—or in the seeming indifference to the suffering of others, as when the poor and homeless, the starving and the marginalized, are left to their own helplessness. Ties persist that bind those who are antagonistic and indifferent to the objects of their antagonism and indifference. The effects of frictions can be far-reaching; hysteresis is at work also in the social context. Moreover, no postmodern fluidity and ephemerality, no dematerialized work of social deconstruction, can deny that there is still something recalcitrantly material in our relations to the world, and something recalcitrantly social even in the apparent dissolution of social ties.

The introduction of a concept like friction into the discourse of the social sciences is likely to meet with resistance and skepticism. To speak about friction in a context other than physics, as the present volume deliberately does, not only entails the crossing of disciplinary boundaries but also goes against a still deeply rooted scientific dualism that separates the world of nature and of technical artifacts from the world of humans, social relationships, and meanings. The separation between the two major scientific cultures persists in multiple guises. It knows prohibitions and taboos. It finds expression in a deep-seated skepticism against transfers and warns against taking natural-science approaches into the social realm.

At stake is not just the jealous guarding of disciplinary fences and professional power-bases. There is the shadow of a long history between the two cultures, memories about the debate on determinism and the possible degrees of freedom for the purposeful shaping of the world through human action that still hang over the cognitive landscape. The long and twisted road of institutionalization of the social sciences has often been accompanied by strategies of deliberate imitation of what was seen to be the secret of the success of the natural sciences: an adherence to the "scientific method," a penchant for quantification and formalization as an alleged safeguard against "ideology," and a striving for what was believed to be a greater degree of "objectivity" to be attained by scrupulously abstaining from too much involvement. On the other hand, each of these arguments could be and were in fact, countered by those who maintained the distinctiveness of the social sciences and humanities. They argued in turn that methods uniquely appropriate for their objects of study had to be found, that no science offered an escape from its social embedding and societal imprinting, and hence that no science was able to escape completely the taint of "ideology." The social sciences in general, and sociology in particular, were shown to oscillate repeatedly in their history between the two opposite poles of attraction: the natural science ideal on one side, with fascination exerted through the discovery of statistical regularities and the search for structural invariants, and literature on the other side, embodying the seduction of the narrative and the interpretation of society as discourse system (Lepenies, 1985). Depending upon the swing of the pendulum, the distance between the respective poles would vary. Undoubtedly, the social sciences (for reasons that must remain undiscussed here) at present lean more toward "the manufacture of words," toward the "reading" of society as a narrative. The fluidity of the postmodern condition does not lend itself well to an empirically minded belief in "facts," even if they are granted to be socially constructed.

Yet there are still other, experimentally induced obstacles litigating against the transfer of metaphors from the natural sciences. In a strict sense, transfers can take place as one of two possibilities: One is the transfer by analogy, entailing either the use of metaphors to stimulate analogous reasoning or the development of mental models or images through analogy. The other form of transfer is much rarer, amounting to a transfer between two different fields of knowledge in the strict sense. In such a case one has to demonstrate how identical mechanisms are at work, whereas in our case only the first form of transfer is intended. But even without claims that isomorphisms or identical mechanisms can be found in the social world corresponding to the natural world, the transfer of metaphors from the natural sciences to the social sciences is fraught with difficulties. Partly as result of the by now discredited transfer of the evolutionary paradigm and its subsequent use and political misuse, especially in the different variants of Social Darwinism, the relationship between the social sciences and biology has remained a tense one ever since. Sociobiology is viewed with suspicion by most social scientists. Reviewing the literature on the Human Genome research recently, Richard Lewontin found ample confirmation for the fear that the dream of mapping the "code of codes" might lead to a renewed genetic determinism as an explanation of all social and individual variation, even if—it could be argued—the genetic code constitutes only a boundary condition for the biological processes of life itself. A medical model of all human variation, Lewontin maintains, makes for a medical model of normality, including social normality, and dictates a therapeutic or preemptive attack on deviance (Lewontin, 1992, p. 34).

The fear of being swept aside by what might become a new model of biological determinism is so deep-rooted that developments in the other sciences are followed only by a small group of social scientists whose job it is to do so: those engaged in social studies of science and technology. For the vast majority of social scientists, developments in the natural and life sciences remain outside their interest and self-restricting competence. As Barbara Adam has shown with respect to the relationship between the concept of physical time and social time, they may go on cheerfully without being aware that their notion of physical time is still based upon concepts borrowed from an outdated and obsolete model of Newtonian physics (Adam, 1990). Worse still, as Philip Mirowski in a scathing and well-founded critique of classical and neoclassical economics recently argued, treating economics as social physics and physics as nature's economics by literally copying equations from thermodynamics has led to serious blunders in the formalization of numerous areas. Thus, neoclassical economics finds itself today in the trap of the "ironies of physics envy" from which escape will not prove easy (Mirowski, 1989).

Borrowing metaphors from the natural sciences has hence remained a risky enterprise for the social sciences. Taken too literally, it might lead into such traps as just mentioned, and practiced "only" on the metaphorical level it might lead nowhere, except into the ups and downs of fads and fashions. Yet, the fateful, if not fatal, attraction persists. The social sciences cannot help "borrowing," just as the natural sciences and their practitioners cannot escape society's cultural grasp on their way of thinking and expressing themselves in language. But what are the legitimate rules for transfers of any kind? How to assess the heuristic fruitfulness, not only of analogies, but of taking them one or two steps further by embedding them in proper theorizing?

In his now-classic study of Weimar physics, Paul Forman (1980) addressed the question of why and how the newly created quantum mechanics, as it emerged in 1925–1926 in a scientific milieu whose center lay in Germany and Austria, was influenced by the general culture surrounding it. In particular, Forman was interested in probing how and why some of the main proponents of the new theory were induced to make public statements and allegations that diverged markedly from what the theory itself warranted. This was especially the case for the concepts of causality, individuality, and intuitive evidentiality or *anschaulichkeit*. In all three cases, quantum mechanics was misused by its main exponents, as Forman claims, to make sweeping epistemic statements that in no way were supported by the theory itself and in some respects were even in outright contradiction to it. Forman concludes that it was due to the intense pressure of the cultural milieu reigning at the time, an anti-intellectual streak of romanticism that celebrated "life" as an unanalyzed experience antithetical to science, a culturally widespread antipathy toward causality, a clinging to the traditional values of individuality, and a popular yearning for *anschaulichkeit* that led to clearing the theory of the stigma of being un-*anschaulich*, causal, and anti-individualistic (Forman, 1971). In his analysis and conclusions, Forman still remains entirely on the conservative scientific side. His analysis is not about physicists' practice in their laboratories, nor about their theories as descriptions of reality. It is meant only as a meta-meta-statement, a statement about the physicists' statements about their description of reality (Forman, 1980). Physicists do not borrow, we might conclude; they merely might get "pressured." Yet, as other examples have also shown, cultural expectations may be so strong that they eventually override correct results.

Subsequent social studies of science and technology have gone much further in deconstructing scientific knowledge claims. No longer seen primarily as descriptions of "reality," they are now deeply implicated by their in-built sociality. Scientific knowledge, and not only meta-statements about it, is susceptible to cultural, political, and economic influences and, as a social construct contingent upon human action and its structuration. Yet, although they go some way in dismantling scientific knowledge claims that partake of a superior cognitive status and access to an "objective" reality, the current streaks of various kinds of constructivism also raise new questions. One of them, of direct interest to those who take a new look at a concept like friction, has to do with the coincidence of worldviews and new scientific theories, of diffuse encounters of an even more diffuse *zeitgeist* and the emergence of new scientific paradigms or paradigmatic shifts. Even if we remain within a social constructivist framework, the precise nature of the constructions, their timing, social acceptance, and inherent robustness still escape us. In Weimar Germany, according to Forman, physicists were swept away to meet implicit and explicit expectations of their general cultural milieu, whether to forestall criticisms or to increase the probability of acceptance of their new theory. Looking across the expert divide, we would also expect the general public or certain "cultural milieus" to be actively on the look-out for scientific confirmations, guidance, and the modicum of certainty that comes with a science-based, cultural worldview. But such general cultural and scientific accommodation is hardly sufficient to explain why at certain times co-evolutionary processes, rooted in very different fields of experience and societal practice, seemingly converge to enable coherence where previously none existed.

For a long time, linearity as a way of seeing the natural world and treating it in mathematical symbols was considered sufficient. Linearity also implied stability. The discovery of the importance of nonlinearity came much later, partly because the mathematics dealing with it lagged behind. It opened up a new world that is inherently instable due to feedbacks and chaos lurking everywhere (Krohn and Küppers, 1989). Friction, with the power to damp instabilities, could therefore no longer be treated only as a nuisance, to be avoided or minimized. Its rediscovery, so to speak, was bound to happen. Once the paradigmatic shift initiated by chaos theory and theories of self-organization had taken hold of the scientific and lay imagination alike, complexity abounded everywhere and so did nonlinearity, critical fluctuations, discontinuities, and variants of chaos ranging from chaotic oscillations to stable chaos found among asteroids in the solar system. The obsession with regularity and order ceased to be of primary interest when new excitement waited for those willing to confront the dynamics of nonlinear systems. And what could be more nonlinear than a social system?

But let us return to the more focused question of when transfers of metaphors from one scientific field of inquiry to another are likely to take place.

One answer may simply be that borrowing is especially attractive whenever we are short of answers, whenever we are in intellectual or "real-world" trouble. Active borrowing is not evenly spread, neither across disciplines nor over time. The motives may vary from imitation to routine, from creative searches for analogies to fashionable trendiness. Yet two observations stand out. One relates to the fact that in each period one scientific discipline or group of disciplines is generally seen to take the lead in intellectual and/or methodological advance. It is thus tempting for others, who see themselves as being less at the forefront of knowledge, to turn toward the more successful scientific fields in search of new orientations. The likelihood that the transfer is actually carried out and accepted is all the greater, the more a kind of successful "coupling" with other, already present strands of theorizing or application can be expected. The virtually instantaneous embracement of Darwinistic evolutionary theory by the social sciences before and around the start of the twentieth century is a case in point. Of course, the theory was adapted in highly selective ways to fit not only already existing theoretical strands but also a much broader political and cultural outlook. This, too, is what happened to the Weimar physicists. The second observation relates to the fact that active borrowing tends to culminate in times when old concepts are felt to be increasingly inadequate to cope with a dramatically changed reality. Borrowing, even if only on the level of analogies, tends to increase when a scientific community is faced with entirely new phenomena for which it has no fitting concepts as yet. When U.S. economists in the last two decades of the nineteenth century were confronted with the partial replacement of the previous "cut-throat competition" among firms and the resulting impact of what was called the "new competition"—namely, the formation of trusts or "combinations" by a few large firms (believed to lead to lower prices and greater efficiency)—they had to rethink older concepts of competition as regulators of the economy and as selector mechanisms for the fit and unfit. The theoretical answers that were given showed great diversity, and there was also great diversity in the nature of evidence, but practically all the borrowing took place from the evolutionary models of the time (Morgan, 1992).

The Janus-Faced Concept of Friction

It is consistent with the persistent inferiority complex of the social sciences that the dramatic changes surrounding the downfall of the communist regimes in Eastern Europe were taken as yet another instance in which they had proven incapable of predicting an event. Rather than greeting the unpredictable as a new phenomenon, as constantly happens in the natural sciences, social scientists reproached themselves almost unanimously for their lack of foresight. And yet the year 1989 was precisely that: an unforeseen and unforeseeable event, the sudden eruption of a much more gradual buildup of severe economic and political deficiencies of a regime that had made believe it was immune to them. That the immediate triggering events occurred, as always, in a historically unique configuration—in this case, with a leadership that avoided outright repression—in no way detracts from the graduality and "hiddenness" of the processes leading to the downfall. This was the year in which friction can be said to have reappeared in the social world as an all-present force on the political scene, precisely at the moment when the bipolar world, with its dynamic equilibrium of terrifying arms, ceased to exist. Though unacknowledged at the time, the vision of a postmodern world society, of an electronic global village, separate from and segregated from other, partly premodern variants (other than the less industrialized Third World countries), came to an abrupt end. Rather, a new and highly explosive mixture of modern, pre- and postmodern elements came to the fore, threatening to impede the dream of a smooth transition to an age of continuing dematerialization where the "symbolic analysts" envisaged by Robert Reich would reign supreme. The countries of Eastern Europe, including those of the former Soviet Union, were too close in geographical proximity and too tied through historical links to the West to be treated with the same mental and political distance that Third World countries receive. Just a few years ago, the capital of Bosnia-Hercegowina, Sarajevo, was under continuous armed attack and thousands of refugees were pouring into the neighboring Western countries. More recently, migratory waves from the East have been triggering highly antagonistic and simplistic responses of xenophobia in Western Europe. Instead of the hoped-for rapid westernization of Eastern Europe in terms of a twin development of market economy and democratization, there are signs of retrograde developments taking place in the West in response to an increase in friction between different levels of modernization and postmodernization. In the United States the effect has not been one linked to geographical proximity; it is of a different kind. The immense difficulties associated with the conversion of the enormous military-industrial complex are a telling example of what can happen when the enemy that mobilized a threat no longer exists.

It is of great importance to realize that friction, in its social as well as physical meaning (since physics after all is employed for human ends), is a janus-faced concept. It acts to dissipate energy that can stabilize or destabilize the interaction. In nonlinear systems, instabilities may occur that open up new pathways of development. Friction as a coupling mechanism bounds these instabilities at finite amplitudes (Küppers, 1993). In social life, friction is implied in every form of interaction, yet too much friction may result in antagonism and violence that entail high costs. A world without friction would be one of complete inertia, without change. One of the few social scientists to deal with this topic, Karl Mannheim, raised the question as to what kind of society would result if the elder generation did not pass away. Friction is a dissipator in the sense that it sweeps away capital accumulated in the social world. This capital may take the form of economic fortunes, empires, and the power accumulated by a few, or, as a biological constant in social life, senescence (i.e., the older generation having to make room for the young). The regulation of friction in such a way that it acts to match energy with what it is to be used for in social life while incurring only gradual losses, rather than sudden and high ones, and yet without preventing change and innovation from occurring, might be considered one of the hallmarks of the civilizing process. Friction in its janus-facedness therefore can be considered to be both a stabilizing and a destabilizing force: It is a stabilizing and a destructive-creative agent unfolding its potential in accordance with specific circumstances. Friction prevents the system from going completely berserk if that system becomes instable—as all nonlinear systems inevitably do. In this instability inherent in nonlinear systems, friction functions as a regulator. It enables the emergence of the new whenever instability reaches a point that no longer can be compensated with old means, while acting as a stabilizer in preventing the system from falling apart completely.

The full recognition of this janus-facedness has far-reaching implications for the social sciences and their approach to the social world. As the children of the Enlightenment that they are, the nascent social sciences have accompanied and carried the process of modernization a long way. As I have shown elsewhere (Nowotny, 1991), the discovery that human intervention could shape societal processes took some time after initially treatment of "the market" and its disruptive consequences in analogy to a natural law—something that could be observed and studied, but not interfered with. The emergence of the modern, secularized nation-state and the firm installment of capitalism in the wake of massive industrialization were extremely powerful forces in shaping Western societies. The social sciences soon became intimately linked with these developments, especially from the latter part of the nineteenth century onward. They not only provided searching interpretations for the deeply disturbing experiences their contemporaries went through, but also offered advice and "solutions" for the social ills that accompanied the processes of industrialization and urbanization. They were passionately engaged in pressing for reforms and new kinds of collective schemes that, after many conflicts, began to take shape in the form of the nascent welfare states. For the administrative buildup, monitoring, expansion, and readjustments of the new administrative and political arrangements of collective health care and education, social security, and transfer payments, the social sciences provided expert scientific knowledge and career prospects but also laid the basis for gradual professionalization (Nowotny, 1991). Moreover, the emergence of expert regimes affected the basic attitudes and outlooks with which people handled their day-to-day interaction and experience: It led to what de Swaan calls protoprofessionalization—that is, to the way in which patients, clients, and schoolchildren but also the public at large "learned" in a simplified and censored version of professional expert knowledge, which was transmitted to them in a continuous formative and informative practice (de Swaan, 1988). It was but a way to prepare them for modernization.

These and other interconnections between the formation and expansion of the modern nation-states, social science knowledge, and the growth of expert regimes were accompanied by still other pressures toward growing interdependence. Standardization was one of the means by which previously uncoupled, idiosyncratic, independently set norms or standards had to become harmonized and standardized in order to permit the buildup of large technological systems. Railroad and electrification necessitated the establishment of large-scale infrastructures, large-scale coordination, and synchronization. Technological systems grew with standardization that went far beyond the unification of purely technical standards encompassing social behavior and practices, the domestication and integration of the hinterland, and the setting up of synchronization and coordination mechanisms, such as world time zones. All these familiar processes working toward standardization and unification, toward order and regularity, predictability and punctuality, assumed new significance when reassessed in terms of friction. The buildup of the welfare state, then, can be reinterpreted as a unique and grandiose attempt to minimize friction, a belated and imperfect attempt to counter the painfully destabilizing aspects of early industrialization and modernization that, indeed, were lived by many contemporaries as a deeply disrupting and disintegrative experience that touched the roots of their social identities and material existence.

The overall emphasis of almost an entire century upon establishing and maintaining order, regularity, and standardization of technology and administration has imperatively pervaded all dimensions of society. Of course, this was a concept of order allegedly derived from nature to legitimate political rule. But it was derived from a thoroughly mechanistic worldview of nature, owing nothing as yet to the concept of self-organization as understood today. The question of political order was a predominant one, not only for social theory but for all of political practice. It was at the basis of the buildup of state bureaucracies and the establishment of an impersonal rule of law. It also formed the basis for the more repressive practices through which the dominant elites attempted to thwart real or perceived threats from the majority of people living within the territorial boundaries of the nation-state who were yet to become integrated as citizens with specified democratic rights and duties, as workers also endowed with social rights, and as mass-consumers of the increasing number and variety of goods and services that the growing and modernizing economy would eventually provide. Regulation and order were the great societal schemes aspired for through such means as teaching hygiene to the poor and "domesticating" them at the same time, instituting obligatory education for all children in preparation of a more predictable, imposed, and orderly worklife whose requirements for values such as regularity, punctuality, and "proper working habits" had to be taught early. These processes of instilling the right attitudes and behavior in workers and citizens went hand in hand with the requirements of massive industrialization. As many historians of technology have shown, the regularity, predictability, and orderliness of the machine was extended to the labor process, and it acted as a powerful metaphor for society at large. The minimization of friction in the orderly running of the machine within the nation-state, which by then was a tightly coupled political and economic system, became the underlying but extremely powerful image that governed the life of men and women as workers and citizens alike. The social costs of setting up and running this machinery were high, as social historians know. On the other hand, destructive energies of enormous proportions were unleashed during the terrible wars among nation-states, with the international system still forming a comparatively fragile, only loosely coupled system to this very day.

The influence of the machine was not limited to the workplace. Yaron Ezrahi has shown convincingly how the rhetoric of the machine and its role as political metaphor has permeated the entire modern liberal-democratic tradition. Science and technology appear to have prepared the ground for "objectifying" political actions as forms of technical public action. The notion that the technicalization of action renders the conduct of actors objectively visible and accountable has encouraged the belief that instrumental accountability—in terms of adequate technical performance— can replace, or at least complement, moral discipline (Ezrahi, 1990, p. 144). Visibility, efficiency, reliability—energy channeled into the directions indicated by planners—included attempts made to minimize friction and gain control over it. What held the machine together were the growing interdependencies of those who served it and were served by it, with friction threatening to reduce its efficiency. All desirable properties of the "political machinery" and the meliorist public policies that were associated with it, especially in the United States, started from the assumption that the performance of the machine could be subject to continuous public improvement. Friction therefore had to be eliminated or at least greatly reduced as standing in the way of any progressivistic policy.

Of course, there were important exceptions to the success in minimizing friction. Even if most nation-states, at least in the industrialized Western world, succeeded within limits in building up internal order and a relatively smooth functioning of the public political sphere, state bureaucracies, and equivalent organizations, even if they succeeded in integrating the masses by extending voting rights to them and by integrating them into the workforce, the minimization of friction between states remained a much more arduous and fragile task. To this day, in a remarkably tragic sense, human history remains a history of war. The "civilizing process" has repeatedly suffered severe breakdowns and returned to barbaric conditions, to the horrors this century is replete with. Wars between states and also within states are still ravaging today. The buildup of international relations and multiply layered ties cutting across borders, although these have undoubtedly progressed in scope and range of preventive and containing mechanisms, remains extremely precarious in view of the mass diffusion and dramatically increased efficiency of technically sophisticated weapons. It is as though a huge potential of friction as a massive trigger of dissipative energies has been implanted in the state construction itself, and is regarded as the ultimate but most efficient response in governing relationships between states. But, at least seen from a Western perspective, ethno-, euro-, or technocentric as it is, such breakdowns and regressions mostly appear as regrettable aberrations, as terrible legacies of the past that must be seen as being balanced by equally remarkable successes in building up economic prosperity and well-being for the majority of citizens. At least the "golden three decades" in the postwar period in Europe, which culminated in an unprecedented growth and expansion of the welfare state, can present themselves as a successful experiment in what—by analogy to the scientific speciality concerned with minimizing friction in engineering—could be called "social tribology," having found the successful sociotechnical means to ease social friction.

In a way, the modernization process, which was begun a century or more earlier, reached its culmination. The evolution of modernity implied what

Norbert Elias had called the establishment of an ever tighter grid of societal coordination and the lengthening of mutual chains of interdependencies. It entailed the establishment of multiple tight coupling mechanisms for which, it turned out, a high price was to be paid in the end. Given a sufficient amount of friction, the instability inherent in any complex nonlinear system—and every social system belongs to this category—would prevent the system from falling apart completely. Friction forestalls complete destruction. But as we have seen, friction was sought to be minimized within the state at least, while it was allowed to erupt time and again between states in the form of war.

Whereas the process of modernization burst into existence with so many painful experiences and the turbulence so eloquently evoked by Marshall Berman (1982) and others, leading to "all that is solid melt(ing) in air," as Karl Marx saw it, modernization, at least provisionally and in one part of the world, converged with other processes that led to a remarkable, if relative, stable internal societal order. It covers an era that can be taken as example, for friction within these societies has become efficiently minimized. Friction can be seen at work in its function as a stabilizer. In order to avoid friction, social consensus increases. Scientific and technological advances played an important role in this process, both in furthering standardization (not to mention "rationalization") and in symbolizing the hope that scientific and technological progress would eventually transform into social progress. Reallocation of wealth, even if it did not occur right away, would become possible tomorrow, since ever more wealth would be available. Such at least was the widely shared assumption that helped to sustain for a long time the belief in progress.

Friction as Dissipative Force: The Rise of Post-Modernity

It is difficult to say precisely when this period came to an end, and when the climax of what was then interpreted as modernization began to merge into postmodern fuzziness. Among the more visible turning points, usually easier to identify in retrospect, were the oil crisis and the incipient awareness of environmental problems to come. Other important changes were under way. The modern—or rather postmodern—information and communication technologies, based upon the successful expansion and application of quantum mechanics in microelectronics and other fields of great technological potential, brought with them different forms of social organization. They expanded into interconnected networks and not, as before, into basically hierarchical forms of organization. The power of computers enabled them to "talk to each other" and to sort out how messages would be packed, and these packages were best routed without passing through the social hierarchical control system every time. In industrial manufacturing, the Fordist type of production process began to be complemented by a "post-Fordist" type, incorporating elements of far greater flexibility, decentralization, and nonhierarchical control. Dematerialization, though grossly exaggerated in the scope already attained, began to fuel the imagination of all those who saw a shift from material production to symbolic production, from goods to services, from the grosser forms of existence to a knowledge-based form. Undoubtedly, the ongoing miniaturization attained especially in microelectronics and the ability to control and manipulate atoms on the microscopic level of matter, as it became feasible in nanotechnology, helped to induce views of an ongoing "dematerialization." Yet, with the shift from the macroscopic level to the microscopic, from hierarchies to heterarchies, from one central and highly visible command post to a multitude of diffuse but highly interlinked network-components, apparently working in decentralized fashion, the degrees of freedom changed accordingly. From a few degrees of freedom on the macroscopic level, many degrees of freedom on the microscopic level resulted. And, in a co-evolutionary societal process, friction, which hitherto had manifested itself mainly on the macroscopic level of society as a social force that had its analogy in mechanics, began to exert itself also on the microscopic level.

Other certainties, until then apparently solid, began to fade with the first serious signs of the erosion of the welfare state. Science and technology, up to now celebrated as the harbingers of technological and economic progress, of the ever-increasing harnessing of the power of nature, began to be viewed with increasing skepticism when it became apparent that continuing scientific and technological advances entailed not just blessings but also risks. Expansion and the human mastery over and exploitation of nature could not go on indefinitely. The limits to growth became a metaphor that pitted those who still believed in linear extrapolation and exponential growth against those who argued for "alternative" paths of development that entailed novel attitudes of restraint, modesty, and even abstention. There were others who based their predictions upon the belief in cyclical developments. Whether these were conceptualized and measured as Kondratieff cycles or in the form of other long-term waves through which societies, the economy, wars, or other phenomena were believed to be passing, the basic idea was a sequence of growth, saturation, and decline, followed by yet another ascension of the curve. In any case, the spell of a world conceived to progress in more or less linear and progressive fashion, thriving upon scientific and technological advances that in turn fueled economic growth and wealth, had been broken.

Environmental threats on a global rather than just local scale began to appear. Not everything that scientific and technological research at the frontier of knowledge declared feasible would be acceptable by society. Large-scale risks, invisible in nature and tendentially global and hence socially indiscriminating, became a source not only of diffuse anxieties but of politically articulate demands for public participation in deciding which kind of future society men and women were striving for. The political use of the metaphor of the machine, so much in tune with the rhetoric of smooth efficiency and instrumentalization of political action, rapidly fell into disrepute. The machine, especially large-scale technology, came to be seen as the villain, an "icon of social excess," as Ezrahi (1990) puts it, no longer associated with discipline and equilibrium. With it went the progressive erosion of the social sciences' influence on both the ideological and intellectual conceptions of society as a causally discernible and manipulative mechanistic system. Society came to be imagined in terms of a narrative about persons, feelings, and values, rather than in terms of causal processes, mechanisms, and mechanistic forces. With this trend went a declining faith in the feasibility not only of monumental engineering in late-twentieth century democracies but also a declining trust in the value of knowledge and rationality in our culture, and a tendency to discredit large-scale schemes of social improvement generally. The result has been the deinstrumentalization of political action leading, in Ezrahi's words, from "statecraft" to "stagecraft" (Ezrahi, 1990, pp. 242–246).

It is precisely amid such widespread tendencies toward fragmentated pluralism, the resurgence of fundamentalism, and, depending upon one's view, that of new social rationalities or irrationalities, amid the loss of previous faith and the relapse into widespread feelings of futility, that the rediscovery of the concept of friction assumes relevance. Friction in past decades functioned primarily as a stabilizer, as an effective control mechanism of social tension and conflicts, permitting more or less deliberate policies that worked as either brakes or accelerators in co-evolutional synchronicity with a polity and society that aspired to the smooth efficiency, instrumentality, and quantitative economic output of the machine. It is only recently that the other side of the janus-faced concept begins to show itself. It is friction as a dissipative force, friction seen at work in its potentially destructive and yet, at the same time, constructive and creative capacity. Friction, instead of functioning mainly as a stabilizer, appears now to be indispensable for any kind of structure to arise and to transform itself. In the well-known Benard experiment, for instance, friction is at work when a fluid is heated up and when spontaneously formed, "self-organizing" patterns begin to emerge. In many cases, friction is the force that takes a system through various states of order by destroying previous ones and leading to the creation of new ones. Its dissipative characteristics are responsible for the loss of energy and its conversion into more "disorderly" states, which in turn, in the process of "self-organization," may generate new forms of "order out of noise."

Friction, as it appears on the microscopic level between atoms displaying its self-organizing effects, becomes an apt metaphor for guidance through present-day changes by focusing upon the self-organizing capacity of societies. On the macroscopic level, friction draws attention through its dissipative capacity to analogies with social forms of dissipation of energy, notably through wars. Among those who employ the concept of friction there is a refusal to accept the loose descriptions of deconstructivism or of descriptions of new societal patterns as everything being "interwoven" with everything else by throwing open the inherent instability of social systems. It opens possibilities for new theorizing in a deliberate nonlinear system mode. Instabilities make for change, but how, when, and in what magnitude are partly regulated through friction. The trap of mere "imitation" of the natural sciences can be avoided, provided that theorizing succeeds in remaining empirically open and sufficiently flexible in combining general features of nonlinear systems dynamics with the empirical particulars of social systems. Friction does not necessarily eliminate the concept of social actor and of agency, as much of systems theory applied to social systems does, for it remains essentially the force that ties together two or more bodies that are in contact with each other— be they fluids or human agents. Moreover, it entails a dynamical dimension, for friction is tied to changes in speed (acceleration or deceleration) and vice versa, thus opening a range of tantalizing questions about this important temporal aspect in society.

Beyond Postmodernism: Living with Friction

Pierre Boulez, the great composer, conductor, and teacher of contemporary music, is reported to have said: "Any musician who has not experienced—I do not say understood, but truly experienced—the necessity of dodecaphonic language is useless. For his entire work brings him up short of the needs of his time." Maybe the lessons that can be drawn from coming to terms with the janus-faced concept of friction are similar: Any social scientist who has not experienced the necessity of friction has failed to understand the exigencies of his or her time. These are exigencies that can no longer be conceptualized in the form of "either/or." The evolution of societies can no longer be cast into multistaged models following an essentially linear path of development with a supposedly more developed, more complex, and usually "superior" stage following the previous one. Modernization may have gone over into postmodernity, but this is no end stage either. A new plurality of cultures and histories of people around the globe are awaiting acknowledgment of both the similarities they share with others and the right to differences that may set them apart. Evolutionary pathways of societies show many bifurcations, periods of stagnation, and breakdowns rather than the smooth and continuously progressive stages that the modernization model presupposed. Global environmental concerns emphasize the urgency of the present human condition. Nature under severe stress in combination with massive human intervention has generated conditions of genuine uncertainty, where nature's resilience at any moment might give way to collapse. A new readiness to accept decentralization and fragmented pluralism as principles of forms of social organization cannot completely neglect the preconditions that guide the search for reaching common agreements. Without a generally accepted frame for agreements and contracts to follow, without, as Durkheim already knew, a setting of precontractual elements for contracts, even a decentralized organization is unlikely to function. Skepticism against any grand scheme, theory, or unifying worldview should not degenerate into cynicism or into a nihilism that denies the validity of any stand taken. The apparent random fluctuations that can be observed in such diverse phenomena as meteorology, wheat prices, and Wall Street share trading should not be taken to mean that order has completely disappeared from social life. Quite the contrary, as the theory of self-organization insists, out of disorder emerges order, and new phenomena of internal coherence appear as the constructive elements of a new world to be constructed: "Such self-determined internal coherences and their natural drift, when observed under contingencies of interactions, will appear as the making of sense, novelty, and unpredictability, in brief, as 'laying down of a world'" (Varela, 1984, p. 30). Naturally, like any construction, this one also knows its own forms of constraints.

Experiencing "the necessity of dodecaphonic language" in the context of social life has indeed been rare and hardly forms part of the professional training or biography of most social scientists today. And yet, at the core of the social sciences, phenomena like social change and innovation have always been central. The former has largely been interpreted as an integral part of the process of modernization, reluctantly and regrettably accepted at times, celebrated at others, whereas the latter has tended to be associated mainly with technology, which, for a long time, was seen as coming into existence in a way external to society. But now the "dodecaphony" of social life is here for all to hear. Far from being without rules, it follows its own; it is not cacophony. Like friction, it is tied to speed and rhythm. Turbulence is likely to arise when, as in a flow of water, with more energy coming into the system, competing rhythms arise, and when each new frequency being introduced becomes incompatible with the preceding one.

That the process of modernization has been closely accompanied by a general increase in the speed with which processes unfold, activities are undertaken, and information, people, and goods are transported and processed has been observed time and again. Acceleration in the scientific-technological domain is contrasted with the much slower rate of change taking place among institutions or people's readiness to "adapt." Intuitively, social scientists have remarked upon differences in speed that characterized separate subsystems and spheres of society, producing various kinds of "lags"; they have also acknowledged the impact of different speeds at which modernization has taken place among different social groups or strata, or across generations. Yet rarely, so far, have these differentials in speed and rates of change been associated with friction or the other way round. Nevertheless, this association is valid and thus permits the opening up of new visions and pathways to fresh analysis and insight. Examples waiting to be studied range from the transition to market economies in Eastern Europe, where speed and sequencing of the various measures and steps in the transformation process certainly are some of the most important variables for success or failure, to the highly unpredictable nature of many nature- and human-induced environmental phenomena under stress. The time and timing of responses delimit the possibilities of human intervention; yet in order to intervene at all, we must know when to do so.

This brings us to a second lesson to be learned from including friction in the social science universe and discourse. Friction does not preclude intervention, it only sets limits on what purposeful action and policy can achieve. To know and study these limits and the conditions under which they occur is to imply the potential improvement of policies. When, to cite but one well-known example: epidemiologists began to study the cyclical nature of epidemics under the aspect of nonlinearity, they found to their surprise that perturbations, like programs of inoculation, initially could lead to huge oscillations. Although any health official or politician who looked at the data would conclude that the program had failed, May and others were able to show that, despite the downward long-term trend, surprising peaks occurred in the short term (May, 1987).

In her comprehensive survey of the influence of natural science theories on contemporary social science, Renate Mayntz (1992) came to the conclusion that it is often the proselytizing efforts on the part of natural scientists, who attempt to generalize their area-specific insights to provide a key to the solution of much more comprehensive problems, that are at the origin of transfers. Social scientists are receptive, because of the evident fact that social phenomena are always irreversible, path dependent, and nonlinear. What attracts them in particular—for example, within the paradigm of self-organization—is the unintentional nature of the processes that allow them to emphasize the autonomous forces of structuration and the unplanned nature of emergent phenomena.They are attracted by physical theories of nonlinear dynamics, as John Casti argues, because they look for suitable conceptual instruments to understand the radical discontinuities that the world around them exhibits, in order to counter them (Casti, 1982). Mayntz provides a sober assessment of the fruitfulness of transfer of methods, concepts, and theories from the natural sciences. The latter transfer, she concludes, practically does not occur. All transfer efforts either remain at the level of verbal analogies—where they fail to carry new information, add nothing to substantive knowledge, and hence limit themselves to a purely semantic innovation—or they are cases of an indirect, mediated theory transfer that proceeds through generalization and respecification. In the latter case, all depends on the social science meaning of an already generalized analytical paradigm and how it is being respecified. Only then can the transfer stimulate a new way of viewing the social phenomena, which, guided by some rather abstract notions, may then trigger a process of social science theory building. She concludes that in a number of cases such stimulating and fruitful impulses for social theorizing have been received, especially when the social science problems have already been relatively well structured. But generally speaking, social scientists have not been compelled to ask entirely new questions (Mayntz, 1992).

Why Should It Be Different for Friction?

The tentative answer I would venture lies in the epistemology of the social sciences having been tightly bound up with the process of modernization and the tutelage of the nation-state. Social friction and the conflicts that went with it were perceived as forces that were to be minimized and that gradually were mastered to the extent that relative stabilization and internal pacification were the result under the constraints described earlier. The present situation no longer allows for such extrapolations from the past. Epistemologies are known to change in accordance with the overall cultural and social contexts in which their objects of study are situated. "Epistemic drifts," as Aant Elzinga called them, occur. The present situation calls for, facilitates, and has already engendered such a change in epistemic outlook. If there is indeed a convergence or, better, a co-evolution taking place between scientific theories and societal development, societies are likely to change the role assigned to friction and their perception of it. As a concept originating in the natural sciences, it will seep into general and social science discourse and increase the readiness to explore disorderly phenomena in their own right and not merely as deviations from an ideal order. Discontinuities, like other social phenomena, become socially real when they are believed to be so. Nongovernability, in comparison with previous experience, becomes an issue when the nation-state is being re-dimensioned everywhere, while the market, though grossly overrated, extends into ever new fields. Among the many unexpected and unpredicted repercussions that recent events in Eastern Europe had, upon the countries immediately involved as well as upon Western Europe, the rediscovery of friction might well be one of the most remarkable and lasting. Civil wars and conflicts over territories in the name of ethnic, national, and linguistic identities not only show the inherent instabilities of the present situation, where even the "falling apart" of a previous nation cannot be achieved without terrible costs, but also raise general questions as to which conditions obtain when coupling mechanisms that link together independent dynamics function reasonably well and give rise to self-organization. Friction is one of these mechanisms. In a similar way the global nature of the ecological crisis supports the concept of friction. Sustainable development will never be reached by planning or establishing a world authority, but only by groping through many levels of action and by motivating millions of human actors. Nature's processes and human intervention have an enormous friction potential—but it need not draw us into endless turbulence.

Innovation of any kind, although Schumpeter meant by it entrepreneurial activity, is indeed part of a process of "creative destruction." There is no reason either to condemn it or to celebrate it, as the social sciences often have done in the past. Attempting to take yet another moral stand will not get us very far. Nor should we fall into the other trap, as has happened often before, and believe that solutions will be forthcoming by subjecting social systems to the rigor with which physical systems can be treated. To include friction in our ways of seeing, theorizing, and communicating about the world entails, first of all, the acknowledgment of its destructive-creative potential. In the inherently instable systems of which we are a living part, both in the social and in the natural world, friction leads us through various states of order in an ongoing process of transformation, while preventing complete destructiveness. Working *with* friction offers an epistemic shift and fresh insights based upon it. Provided that the concept of friction remains empirically open and rooted in specific fields of inquiry, it is sufficiently rich to lead to new questions in the social sciences that have a practical interest built into them in addition to the theoretical one. Perhaps most important of all, friction offers a perspective to go beyond postmodernity in trying to make sense of the dodecaphonic (or however many) voices around us. Not everything must, nor will, end in the deconstruction of sense, texts, and the world at large. Working with friction also entails readiness to encounter surprises and to meet the unpredictable in what is, hopefully, a prepared way. There is a lot to be said for experiencing friction—as a prescientific immersion into a widespread collective experience—in its destructive-creative potential in order to try to understand it better. Learning how to live with friction, while seeing its janus-face in a holographic perspective and not forgetting the potential it holds out for controlling and utilizing it, may after all be what is needed to guide us through postmodernity.

References

Adam, Barbara. *Time and Social Theory*. Oxford: Polity Press, 1990.
Berman, Marshall. *All That Is Solid Melts into Air: The Experience of Modernity*. New York: Simon and Schuster, 1982.
Casti, John. "Topological Methods for Social and Behavioral Systems," *International Journal of General Systems*, 8, 187–210, 1982.
de Swaan, Abram. *In Care of the State. Health Care: Education and Welfare in Europe and the USA in the Modern Era*. Cambridge: Polity Press, 1988.
Esping-Andersen, Gösta. "Post-Industrial Class Structures: An Analytical Framework." Madrid: Instituto Juan March de Estudios e Investigaciones, Working Paper 1992, No. 38, 1–35.
Ezrahi, Yaron. *The Descent of Icarus: Science and the Transformation of Contemporary Democracy*. Cambridge, Mass.: Harvard University Press, 1990.
Forman, Paul. "Weimar Culture, Causality, and Quantum Theory, 1918–1927: Adaptation by German Physicists and Mathematicians to a Hostile Intellectual Environment." In Russell McCormmach (Eds.), *Historical Studies in the Physical Sciences*. Philadelphia: University of Pennsylvania Press, 1971.

————. "Kausalitèt, Anschaulichkeit und Individualitèt, oder wie Wesen und Thesen, die der Quantenmechanik zugeschrieben, durch kulturelle Werte vorgeschrieben wurden" (pp. 393–406). In Nico Stehr und Volker Meja (Eds.), *Wissenssoziologie: Sonderheft 22 der Kölner Zeitschrift für Soziologie und Sozialpsychologie*. Opladen: Westdeutscher Verlag, 1980.

Harvey, David. *The Condition of Postmodernity*. Cambridge: Blackwell, 1990.

Krohn, Wolfgang, and Küppers, Günther. *Rekursives Durcheinander* (pp. 69–83). Berlin: Kursbuch 98, 1989.

Küppers, Günther. *Social order*. 1993.

Lepenies, Wolf. *Die drei Kulturen: Soziologie zwischen Literatur und Wissenschaft*. München: Carl Hanser, 1985.

Lewontin, Richard C. "The Dream of the Human Genome," *New York Review of Books*, May, 1987, 31–40.

————. "Nonlinearities and Complex Behavior in Simple Ecological and Epidemiological Models," *Annals of the New York Academy of Sciences*, 504, July 1–15, 1992.

Mayntz, Renate. "The Influence of Natural Science Theories on Contemporary Social Science." In M. Dierkes and B. Biervert (Eds.), *European Social Science in Transition: An Assessment and Outlook*. Frankfurt am Main; Campus, 1992.

Mirowski, Philip. *More Heat Than Light: Economics as Social Physics, Physics as Nature's Economics*. Cambridge: Cambridge University Press, 1989.

Morgan, Mary S. "Competing Notions of 'Competition' in Late-Nineteenth-Century American Economics." Paper presented at the Conference on "Transfer of Metaphors Between Biology and the Social Sciences," Bielefeld, June 22–24, 1992, pp. 1–49.

Nowotny, Helga. "Knowledge for Certainty: Poverty, Welfare Institutions and the Institutionalization of Social Science" (pp. 23–44). In P. Wagner and B.Wittrock (Eds.), *Discourses on Society: Yearbook in the Sociology of the Sciences*, Vol. 15. Dordrecht: Kluwer Academic Publishers, 1991.

Reich, Robert B. *The Work of Nations: Preparing Ourselves for 21st Century Capitalism*. New York: Alfred Knopf, 1991.

Schuller, T. "After Employment: Unemployment, Retirement, Time and Gender." *Ms.*, 1988.

Notes

Chapter 2

1. For an example of a study that stresses the importance of psychological consideration when choosing alternative means of planning artillery support, see the Soviet military monthly *Voenny Vestnik*, No.4 (1980), p. 64. The author, drawing from Soviet exercises, personal experience, and the lessons of the Yom Kippur War, concludes that the success of a high-speed offensive depends on the total temporary loss of combat effectiveness by the enemy. He gives the following as recovery times for defenders in trenches 5–10 meters from their weapons and subjected to accurate bombardment for six minutes. In other words, these are the times taken to open fire after the last Soviet round has fallen on the position:

machine guns	—	45 seconds
antitank guns	—	2-3 minutes
ATGWs (Milan-type)	—	60–72 seconds
tanks	—	2–3 minutes

Chapter 4

1. According to J. Viner (1978), "This Protestant 'ethic' resulted, at least for the lesser ranks of the entrepreneurial class in dedicated and unlimited pursuit of wealth through unremitting industry, rigid limitation of expenditures on personal consumption or charity, concentration of time and attention on the pursuit of one's business affairs, avoidance of distraction through intimate friendship with others, systematic and pitiless exploitation of labor, and strict observance of honesty in one's relations with others within limits set by 'formal legality.' The sole concern of the ethic was the service of the Glory of God moved by inscrutable objectives; any social consequences would be a fortuitous by product" (pp. 151–152). And as R. H. Tawney pointed out, "Religion influenced men's outlook on society to a degree which today. . . is difficult to appreciate. Economic and social changes in turn acted powerfully on religion. Weber emphasized only the first point."

Chapter 5

I am grateful to Nordal Åkerman for his comments on an earlier version of this chapter. They have helped improve the final product.

Chapter 6

I am grateful to my colleagues at the University of California at Riverside for their helpful comments on an earlier version of this chapter: Susan Carter, Stephen Cullenberg, Gary Dymski, David Fairris, Robert Pollin, and Carl Uhr.

1. See Nicholas Kaldor, "The Irrelevance of Equilibrium Economics," *Economic Journal*, December 1972.

2. See H. A. Simon, "A Behavioral Model of Rational Choice," *Quarterly Journal of Economics*, Vol. 59, 1955.

3. The pioneering paper on rational expectations is John F. Muth's, "Rational Expectations and the Theory of Price Movements," *Econometrica*, July 1961.

4. However, there can be "real" business cycles caused by exogenous, random fluctuations in the pace of technological change. See John B. Long, Jr., and Charles I. Plosser, "Real Business Cycles," *Journal of Political Economy*, February 1983.

5. Robert Solow, "Economic History and Economics," *American Economic Review*, May 1985, p. 328.

6. Joan Robinson scathingly criticizes this way of thinking, saying that "in time, the distance between today and tomorrow is twenty-four hours forwards, and the distance between today and yesterday is eternity backwards." See her "Lecture at Oxford by a Cambridge Economist," in her *Contributions to Modern Economics*, 1978, p. 139. Also see her "History Versus Equilibrium" in the same volume.

7. See Paul David, "Path-Dependence: Putting the Past into the Future of Economics," IMSSS Technical Report No. 533, Stanford University, 1988; W. Brian Arthur, "Self-Reinforcing Mechanisms in Economics," in Phillip N. Anderson, Kenneth J. Arrow and David Pines, eds., *The Economy as an Evolving Complex System*, Addison-Wesley, 1988; Paul David, "Clio and the Economics of QWERTY," *American Economic Review*, May 1985.

8. Paul Krugman, "History and Industry Location: The Case of the Manufacturing Belt," *American Economic Review*, May 1991.

9. The example that follows is taken from Keith Griffin, *Alternative Strategies for Economic Development*, 1989, p. 44.

10. For an analysis of fixed prices within a general equilibrium framework see Robert Barro and Herschel Grossman, "A General Disequilibrium Model of Income and Employment," *American Economic Review*, March 1971.

11. See, for example, Ronald Coase, "The Nature of the Firm," *Economica*, November 1937. For a more modern statement see Oliver E. Williamson, "Transactions-Cost Economics: The Governance of Contractual Relations," *Journal of Law and Economics*, Vol. 22, 1979.

12. See Gregory Mankiw, "Small Menu Costs and Large Business Cycles: A Macroeconomic Model," *Quarterly Journal of Economics*, May 1985; George Akerlof and Janet Yellen, "A Near-Rational Model of the Business Cycle, with Wage and Price Inertia," *Quarterly Journal of Economics*, Supplement, 1985.

13. Russell Cooper and Andrew John, "Coordinating Coordination Failures in Keynesian Models," *Quarterly Journal of Economics*, August 1988.

14. Laurence Ball and David Romer, "Sticky Prices as Coordination Failure," *American Economic Review*, June 1991.

15. Susan B. Carter and Elizabeth Savoca, "Labor Mobility and Lengthy Jobs in Nineteenth-Century America," *Journal of Economic History*, March 1990; Susan B. Carter and Richard Sutch, "The Labor market in the 1890s: Evidence from Connecticut Manufacturing," in Barry Eichengreen, ed., *Underemployment and Unemployment in Historical Perspective*, Leuven University Press, 1990.

16. Harvey Leibenstein, "The Theory of Underemployment in Backward Economies," *Journal of Political Economy*, April 1957; Joseph Stiglitz, "The Efficiency Wage Hypothesis, Surplus Labor, and the Distribution of Income in L.D.C.s," *Oxford Economic Papers*, July 1976.

17. See, for example, Janet Yellen, "Efficiency Wage Models of Unemployment," *American Economic Review*, May 1984.

18. Laurence Ball and David Romer, "Real Rigidities and the Non-Neutrality of Money," *Review of Economic Studies*, April 1990.

19. Perhaps a technical point is in order here. The fact that a market has not cleared does not necessarily imply that the market is in disequilibrium. It is possible to conceive of markets which are in equilibrium, i.e., in which there is no tendency for price or quantity to alter, yet which none the less have failed to clear. Most economists would regard such a situation as exceptional however.

20. Joseph Stiglitz and A. Weiss, "Credit Rationing in Markets with Imperfect Information," *American Economic Review*, June 1981.

21. This view is put forward in Jo Anna Gray and Magda Kandil, "Is Price Flexibility Stabilizing? A Broader Perspective," *Journal of Money, Credit and Banking*, February 1991.

22. For a formal model that yields this conclusion see J. Bradford De Long and Laurence H. Summers, "Is Increased Price Flexibility Stabilizing?", *American Economic Review*, December 1986.

23. *Ibid.*, pp. 1042-3.

24. My colleague Carl Uhr commented that the problems created by friction can perhaps best be overcome by collectivizing them, as in the development of social security systems and safety nets against poverty. Government action in such a view is not a source of friction but a response to friction.

Chapter 7

The author wishes to thank the John Simon Guggenheim Memorial Foundation for generously providing the Fellowship during which most of the substance of this chapter was developed. Research in high-reliability organizations was partially supported by the Office of Naval Research and by the National Science Foundation.

1. Taylor referred to this phenomenon as "soldiering." In later times, labor adopted the refusal to work beyond the minimum requirements as a strategy with the somewhat more benign label of "work to rule." The implication is that the rules remain minimal rather than typical.

2. Somewhat ironically, Taylor's first major inquiry concerned the increase of machine tool speeds. In fact, the discovery of high-speed cutting steels that could withstand the frictional heating without losing their edge was his first major accomplishment. See, for example, Merkle (1980), p. 11.

3. The term *scientific management* is rarely used by those who declare their field to be management science. They sharply contrast their use of models of limited rationality in the open or natural systems approach with the formal, closed-system perspectives of the scientific management school. Nevertheless, the modern focus on "optimal" solutions seems quite reminiscent of the Taylorist approach.

4. Possible examples of potential direct harm include the consequences of automating air traffic control. Indirect harm, on the other hand, might arise from, say, fully computerizing and automating credit records or police files.

5. The extensive literature on risk and risk acceptance is too large to summarize here. However, three useful sources are American Academy of Arts and Sciences (1990), Douglas and Wildavsky (1983), and Kates, Hohenemser, and Kasperson, (1985).

6. Examples, rarely analyzed in the literature, include the General Electric nuclear fuel reprocessing plant in Tennessee, which was closed down on safety grounds before it ever opened, and the Rancho Seco nuclear reactor in California.

7. This category includes even those cases where fault is found with inadequate training in general when they deny the possibility that no amount of training could have guaranteed a better outcome. See, for example, Rochlin (1991).

8. This example is less applicable to Europe, where the craft guild and apprentice system survives in many pockets of even high-technology industries.

9. For example, a television series on artificial intelligence described a computer-based medical diagnostic program, which, fed with a description of an old Chevrolet with rust spots, diagnosed it as having measles. The assumption that the entity being diagnosed was a human being, and alive, was programmed in.

Chapter 12

1. Sigfried Giedion, *Space, Time and Architecture* (1941), part III.

2. Perhaps the clearest documentation of this change of attitude is to be found in Gunnar Asplund et al., *Acceptera* (Stockholm, 1931). The creators of functionalism were aware of the political consequences of the new architecture as can clearly be seen in the final chapter of Le Corbusier's *Vers Une Architecture* (Toward a New Architecture) (1923; reprinted in 1946). This chapter is entitled "Architecture or Revolution."

3. Kaj Nyman, *Husens språk* (The language of buildings) (Stockholm, 1989), ch. 2.

4. What survives in nature, "in the long range, is not the fittest individual, organism, population or species, but the fittest eco-system: *both* system *and* environment." See Anthony Wilden, *The Rules Are No Game* (London, 1987), p. 192.

5. My use of the words *human, man,* and *people* subjects the reader to the risk (as Wilden says) of "confusing society, a product of history, with the species, a product of natural evolution" (Ibid., p. 71). I urge caution, therefore, every time these words appear in the text.

6. This argument is based mainly upon Wilden's, *System and Structure* (London, 1980).

7. "Without context there can be no. . . . Communication is impossible without coding and coding is impossible without context" (Wilden, *The Rules Are No Game,* p. 181).

8. This comment comes originally from Konrad Lorenz.

9. See Gregory Bateson, *Steps to an Ecology of Mind* (New York, 1973).

10. This is the so-called Sapir-Whorf hypothesis.

11. Nyman, *Husens språk,* ch. 4.

12. Cf. Heidegger.

13. Reconstruction of housing conditions is one of the very best sources of anthropological knowledge.

14. A related point is this: "To restore both inertia and the possibility of unanticipated events—that is, restore the open character of history—we must accept its fundamental uncertainty. Here we could use as a symbol the apparently accidental character of the great cretaceous extinction that cleared the path for the development of mammals, a small group of ratlike creatures." See Ilya Prigogine and Isabelle Stengers, *Order Out of Chaos* (London, 1984), p. 207.

15. Ibid., pp. 188ff.

16. As Wilden points out in *System and Structure*: "Complexity and manifoldness are a result of restrictions" (p. 170). Did not Ludwig Wittgenstein make the same point in *Tractatus*? "The limits of my language mean the limits of my world. . . . If the good or bad exercise of the will does alter the world, it can alter only the limits of the world. . . . The effect must be that it becomes an altogether different world." See Wittgenstein, *Tractatus Logico-Philosophicus* (London, 1922; rev. trans. 1961), sections 5.6 and 6.43.

17. In a similar vein, Sartre noted that "man constructs signs because in his very reality he is signifying; and he is signifying because he is a dialectical surpassing of all that is simply given. What we call freedom is the irreducibility of the cultural order to the natural order." See Sartre, *Question de Méthode* (Search for a Method) (Paris, 1960; reprinted in 1963), p. 152.

18. In this connection, see Hazel Barnes's interpretation of Sartre's humanism in the Foreword to *Questions de Méthode.*

Chapter 14

1. Jay Doblin, *Discrimination: The Special Skills Required for Seeing, and the Curious Structure of Judgment* (privately printed by Jay Doblin, Keeley, Malin, and Stamos, Chicago, 1990).

2. See Matthew Turner, *Made in Hong Kong: A History of Export Design in Hong Kong* (Hong Kong Urban Council, 1988).

3. Alison Kelly, *The Story of Wedgwood*. (London: Faber and Faber, 1930; revised in 1975), p. 34.

4. W. Cooke Taylor, "Art and Manufacture," *Art-Union*, March 1, 1848; quoted in C. Harvie et al., Eds., *Industrialization and Culture 1830–1914* (London: Macmillan/Open University Press), 1970.

5. "Philistines Lose the Battle of the Boxes," *Sunday Times*, July 7, 1988.

6. Ibid.

7. Ibid.

8. "British Telecom to Review Its Identity," *The Independent*, July, 28, 1989.

9. Alfred P. Sloan, Jr., *My Years with General Motors* (New York: Doubleday, 1963; reprinted in 1990). p. 264.

10. Ibid.

11. "They're Still Groping: Just What Is the Right Image for Oldsmobile?," *Business Week*, July 9, 1990.

Chapter 17

1. Charles E. Lindblom, "The Science of 'Muddling Through,'" *Public Administration Review,* 19, 1959, 79-88.

2. Jay W. Forrester, "Lessons from System Dynamics Modeling," *System Dynamics Review*, 3, no. 2, Summer 1987, 136-149.

3. Paul J. Werbos, "A Menu of Designs for Reinforcement Learning over Time," in *Neural Networks for Control*, T. Miller, R. Sutton., and P. Werbos, Eds. (Cambridge, Mass.: MIT Press, 1990), pp. 67-95.

4. Herbert A. Simon (with W. G. Chase), "The Mind's Eye in Chess," in *Models of Thought*, Herbert A. Simon, Ed. (New Haven, Conn.: Yale University Press, 1979), p. 421. See also Simon's Hitchcock Lecture delivered at the University of California at Berkeley, February 13, 1990.

5. A more detailed exposition of our skill-acquisition model, directed toward coping skills to a greater extent than toward predictive skills, may be found in Hubert L. Dreyfus, and Stuart E. Dreyfus, *Mind over Machine* (New York: Free Press, 1988).

6. Dreyfus and Dreyfus, *Mind over Machine*, pp. 36-40, 163-167.

About the Authors

Åke E. Andersson is a Professor of Economics and Managing Director of the Swedish Institute for Futures Studies in Stockholm. He was awarded the Honda prize in 1995. His recent publications in English are *Advances in Spatial Theory and Dynamics* (1989), *Knowledge and Industrial Organization* (1989), *The Complexity of Creativity* (1997), and *Government for the Future* (1997).

Ottar Brox is a Professor of Sociology and Research Director of the Norwegian Institute for Urban and Regional Research in Oslo. He served as a member of the Norwegian Parliament from 1973 to 1977. His books include *Newfoundland Fishermen in the Age of Industry* (1972) and *"Jeg er ikke rasist, men..."* ["I am not a racist, but..." How do we develop attitudes to immigration and immigrants?] (1991).

Sigrid Combüchen is a novelist, critic, and essayist. One of her recent books is *Byron: A novel* (1988), published in nine languages (e.g., in the U.K. in 1991).

Chris Donnelly is a Special Adviser on Central and East European Affairs to the Secretary General of NATO and Director of Soviet Studies Research Center at the Royal Military Academy in Sandhurst. He is author of *Red Banner: The Soviet Military System in War and Peace* (1988). and editor of *Gorbachev's Revolution* (1989).

Hubert L. Dreyfus is a Professor of Philosophy at the University of California, Berkeley. He specializes in phenomonology and contemporary philosophy. His books include *Mind Over Machine* (with Stuart L. Dreyfus, 1986), *What Computers Still Can't Do* (1992), and *Heidegger* (1992).

Stuart E. Dreyfus is a Professor of Industrial Engineering and Operations Research at the University of California, Berkeley. He specializes in artificial neural systems, cognitive ergonomics, and optimal control theory. His publications include *The Art and Theory of Dynamic Programming* (1977) and *Mind Over Machine* (with Hubert L. Dreyfus, 1986).

Jon Elster is a Professor of Political Science and Philosophy at the University of Chicago. His publications include *Ulysses and the Sirens* (1979), *Solomonic Judgements* (1989), *Local Justice* (1992), and *Political Psychology* (1993).

Joanne Finkelstein is a social theorist. She has lectured at American universities and currently teaches at Monash University, Australia. She is the author of *Dining Out: An Observation of Modern Manners* (1989), *The Fashioned Self* (1991) and *After a Fashion* (1996).

Keith Griffin is a Professor and Chairman of the Department of Economics, at the University of California, Riverside. He was formerly president of Magdalen College, Oxford. Among his books are *Alternative Strategies of Economic Development* (1989) and *Studies in Globalization and Economic Transitions* (1996).

Rom Harré is a Professor of Psychology at Georgtown University, Washington D. C. and a Fellow of Linacre College, Oxford. His publications include *The Principles of Scientific Thinking* (1970), *Causal Powers* (1975), *The Social Construction of Emotions* (1986), and *The Principles of Linguistic Philosophy* (1997).

Agnes Heller is a Hannah Arendt Professor of Philosophy, at the New School of Social Research, New York. Earlier, she was Associate Professor at the University of Budapest, then lecturer of sociology in Melbourne. Among her works are *Everyday Life* (1984), *The Power of Shame* (1985), *Can Modernity Survive?* (1990), and *An Ethics of Personality* (1996).

Jon Heskett is a Professor of Design at the Institute of Design at Illinois Institute of Technology, Chicago. He is a consultant to organizations around the world and has authored *Industrial Design* (1979), *Design in Germany 1870–1918* (1986), and *Philips; A Study in Corporate Management of Design*, (1989).

T. R. Lakshmanan is a Professor of Geography and Director of the Center for Energy and Environmental Studies at Boston University, Boston. Among his books are *Urbanisation and Environmental Quality* (1977), *Systems and Models for Energy and Environmental Analyses* (1983), *Large Scale Energy Projects* (1985), and *Structure and Change in the Space Economy* (1993).

Helga Nowotny is a Professor and head of the Institute for Theory and Social Studies of Science, University of Vienna. She was the Founding Director of the European Center in Vienna from 1974 to 1986 and Chairperson of the Standing Committee for the Social Sciences of the European Science Foundation from 1985 to 1991. Her recent works are *Self-Organization—Portrait of a Scientific Revolution* (with W. Krohn and G. Küppers, 1990), *In Search of Usable Knowledge* (1990), *The Sociology of the Sciences* (1996), and *After the Breakthrough* (1997).

Kaj Nyman is a Professor of Urban Planning at the University of Uleåborg, Finland, as well as an architect. He authored *Husens språk* [The Language of Buildings] in 1990.

Klaus Rifbjerg is a novelist, poet, critic, essayist, and filmmaker. He served as Literary Director of Gyldendal Publishers in Copenhagen from 1985 to 1991. Some translated books of his are *Selected Poems* (1976), *Anna (I) Anna* (1982), *Witness to the Future* (1984), and *Afterbeat* (1997).